Risk Assessment and Decision Making Using Test Results

The Carcinogenicity Prediction and Battery Selection Approach

Risk Assessment and Decision Making Using Test Results

The Carcinogenicity Prediction and Battery Selection Approach

Julia Pet-Edwards and Yacov Y. Haimes
University of Virginia
Charlottesville, Virginia

Vira Chankong, Herbert S. Rosenkranz, and Fanny K. Ennever
Case Western Reserve University
Cleveland, Ohio

PLENUM PRESS • NEW YORK AND LONDON

Library of Congress Cataloging in Publication Data

Risk assessment and decision making using test results: the carcinogenicity prediction and battery selection approach / Julia Pet-Edwards . . . [et al.].
   p.   cm.
Includes bibliographies and index.
ISBN 0-306-43067-3
1. Carcinogenicity testing—Decision making. 2. Health risk assessment—Decision making. I. Pet-Edwards, Julia.
RC268.65.R57   1989   89-3499
616.99'4071—dc19   CIP

© 1989 Plenum Press, New York
A Division of Plenum Publishing Corporation
233 Spring Street, New York, N.Y. 10013

All rights reserved

No part of this book may be reproduced, stored in a retrieval system, or transmitted in any form or by any means, electronic, mechanical, photocopying, microfilming, recording, or otherwise, without written permission from the Publisher

Printed in the United States of America

Preface

The difficulties associated with making risk assessments on the basis of experimental results are familiar to practitioners in many fields—engineering, epidemiology, chemistry, etc. These difficulties are particularly common in problems that have dynamic and stochastic characteristics driven by multiple purposes and goals, with complex interconnections and interdependencies. Acquiring an appropriate data base, processing and analyzing model results, and transmitting these results at an appropriate technical, social, political, and institutional level are additional difficulties that must be addressed.

This book is grounded on the premise that risks are best assessed on the basis of experimental results and sound mathematical analyses, coupled with the knowledge of experts. The carcinogenicity prediction and battery selection (CPBS) approach described herein provides a systematic mechanism—a synthesis of systems and statistical and decision analyses—to aid researchers and decision makers in the critical field of carcinogenicity prediction in selecting an appropriate battery of tests to use and in translating experimental results into information that can be used as an aid to decision making.

The book is divided into two parts. Part I provides most of the mathematical and systems engineering background needed to understand the CPBS approach. Part II provides the details of the CPBS approach and concludes with a case study involving the use of the CPBS for cancer hazard identification. Example problems have been added throughout the book to illustrate uses of the methodologies and techniques, and to enhance pedagogy, comprehension, and learning. The theoretical basis that provides the scientific and mathematical support for the CPBS approach is kept to a minimum. This book is within the realm of readers who have an understanding of elementary concepts of probability and statistics. A background in the basic concepts of optimization would be useful but is not a necessary prerequisite for reading this book. Ample references to the available literature address the interests of the more theoretically oriented reader.

Risk assessment of chemical carcinogens is in the domain of many federal, state, and local regulatory agencies as well as large and small chemical and pharmaceutical companies. This book will be an invaluable source of information on how to evaluate the potential carcinogenicity of chemicals and how to design the best battery of short-term tests for this purpose. Although the CPBS approach was initially developed to address the problem of how to utilize short-term tests in screening for potential chemical carcinogens, we have found that this unique blend of methodologies has a much broader potential for use. For example, this approach has been useful in medical diagnosis, fault diagnosis, and in the identification of environmental hazards. Consequently, this book is written for college students and researchers, as well as for engineers, managers, regulatory agencies, and other professionals involved in the problems of assessing risks on the basis of multiple test results.

The authors would like to thank R. Galen, G. Klopman, and K. Loparo for their constructive comments and suggestions and Virginia Benade for her editorial assistance; and C. J. Debeljak, J. Mitsiopolis, and J. Donahue for their assistance in analyzing the Gene-Tox data base. Special thanks should be extended to the National Science Foundation, Charles A. Dana Foundation, Case Western Reserve University Medical School, and NASA Lewis Center for supporting some of the research documented in this book. However, the authors and not the above agencies and institutions are responsible for the results and ideas presented here. The authors would further like to acknowledge *Mutation Research* for granting copyright permission to use material from published papers. Finally, we would like to thank Case Western Reserve University and the University of Virginia for their overall support; and we are particularly indebted to our colleagues and students whose advice, suggestions, and comments helped us to bring this volume to its present form.

Julia Pet-Edwards
Yacov Y. Haimes
Vira Chankong
Herbert S. Rosenkranz
Fanny K. Ennever

Charlottesville and Cleveland

Contents

Part I	Basics of the CPBS Approach to Risk Assessment and Decision Making	1
Chapter 1	The Carcinogenicity Prediction and Battery Selection Approach	3
	1.1. Introduction	3
	1.2. Overview of the CPBS Methodology............	5
	1.3. The Cancer Hazard Identification Problem	8
	References	13
Chapter 2	Fundamental Basics of the CPBS Approach	15
	2.1. Bayesian Decision Analysis.....................	17
	2.2. Cluster Analysis	31
	2.3. Multiple-Objective Decision Making.............	43
	2.4. Dynamic Programming	53
	References	62
Part II	Carcinogenicity Prediction and Battery Selection Methodology.................................	65
Chapter 3	Preliminary Analysis..............................	67
	3.1. Data Base Considerations	70
	3.2. Analysis of Test Performance	74

	3.3. Exploratory Analysis of the Data	85
	3.4. Testing for Dependencies among Tests	100
	References	122
Chapter 4	Battery Selection	125
	4.1. Computing Measures of Battery Performance	127
	4.2. Battery Selection Using Enumeration	138
	4.3. A Heuristic Approach for Constructing Batteries with Good Performance	142
	4.4. A Strategy for Constructing Batteries Using Dynamic Programming	148
	4.5. Dynamic Programming for Construction of Non-dominated Batteries	155
	References	164
Chapter 5	Risk Assessment Using Test Results	165
	5.1. Making Predictions When Tests Are Conditionally Independent	166
	5.2. Making Predictions When Tests Are Conditionally Dependent	170
	Reference	173
Chapter 6	Applications of CPBS to Cancer Hazard Identification	175
	6.1. Introduction	175
	6.2. An Illustration of CPBS Prediction on Acrylamide	180
	6.3. Battery Selection Considerations	184
	6.4. CPBS and Decision Analysis	194
	6.5. Closing Remarks	199
	References	200
Epilogue		203
Appendix. List of Abbreviations for Bioassays and Short-Term Tests		207
Index		209

Part I

Basics of the CPBS Approach to Risk Assessment and Decision Making

This book is concerned with the problems in decision making that occur when multiple test results are available. The carcinogenicity prediction and battery selection (CPBS) approach is a collection of methods that we have found useful in aiding researchers and industrial decision makers in selecting the best set of tests to use for a given problem and interpreting the results of a battery of tests. Part I consists of two chapters in which we discuss the basics of the CPBS approach. In Chapter 1 we discuss the underlying philosophy behind the CPBS approach and briefly outline its major components. We follow this with a description of the cancer hazard identification problem—the genesis of the CPBS approach, which also provides the theme for the majority of the examples used to illustrate basic concepts and procedures throughout this book. In Chapter 2 we discuss four different methodologies and decision tools—Bayesian decision analysis, cluster analysis, multiple-objective decision making, and dynamic programming—which form the foundations of the CPBS approach.

Chapter 1

The Carcinogenicity Prediction and Battery Selection Approach

1.1. INTRODUCTION

Decisions are most often based upon the results of experiments coupled with the knowledge of experts. When sampling or experimental results are available, they often constitute the major *factual* information input into the decision-making process. For example, clinicians and physicians use diagnostic tests and clinical findings along with their expert knowledge to diagnose their patients' problems. When the physician estimates that the "risk" of a disease is high enough (here we define "risk" as both the probability and the severity), expert knowledge is again used to develop appropriate treatment plans. Toxicologists (in industry, government, and academia) use test results on live animals as well as short-term *in vitro* tests to study the carcinogenic potential of chemicals. If the risk of carcinogenicity for a particular chemical is high, then a pharmaceutical or chemical company may decide to stop or delay the development of the chemical, a regulatory agency may decide to ban the development or restrict the use of such a chemical, or a researcher in academia may decide to study this chemical further to examine its modes of action. In industry, quality control managers interpret results from multiple "inspectors" (humans or machines) to identify defective parts and to decide whether a defective part should be destroyed or sent to a rework station. Water resources and environmental engineers utilize results from well sampling to decide what type of action is warranted on an aquifer found to be contaminated. These are but a few examples indicating how often experimental results enter into decisions in a diversity of problems.

This book is about decision making when a choice of multiple test results is available. The carcinogenicity prediction and battery selection (CPBS) approach to industrial decision making is a collection of methodologies that translates knowledge concerning the performances of multiple tests into a form that can be used to improve decisions. The approach is grounded on the premise that decisions should be based upon sound scientific analyses of test results as well as the judgment of experts. In the context of the cancer hazard identification problem, the analysis of the test results on a particular chemical leads to a description of the likelihood (probability) that the chemical is carcinogenic. Then the actual decision as to whether the likelihood of carcinogenicity is strong enough to warrant action is left to the decision maker.

Most tests and measuring devices are imperfect. They often produce two types of errors: random errors (due to the imprecision of the test procedures or measuring instruments) and systematic errors or bias (due to the inaccuracy of the tests). Random errors can be detected and quantified through the repeated use of a single test. However, with systematic errors, a single repeated test may not be adequate; consequently, the use of more than one test or experiment may be desirable. The difficulties involved with the use of multiple tests in decision making are twofold: (1) how to select from all available tests and experiments a battery of tests most suited to the problem at hand, and (2) how to interpret the (often mixed) results of a battery of tests.

These two problems can be stated formally as follows:

Battery Selection Problem. Given: (1) a set of n tests, (2) previous results of these tests on a set of m_1 objects (e.g., patients, chemicals, parts) having the property of interest (e.g., disease, carcinogenicity, defect) and m_2 objects without the property, and (3) "costs" associated with each test (e.g., dollars, time, difficulty). Then the goal of the battery selection process is to make use of the information on each individual test in order to formulate a strategy for finding the "best" battery of k tests—best in terms of collective performance, minimum testing time, minimum testing costs, and risk behavior of the decision maker, or a compromise among these attributes. Note that the number of tests may be fixed *a priori* or may be considered a decision variable.

Prediction Problem. Given: (1) an object with unknown property, (2) the results of a set of k tests on the given object, and (3) the past performance of these tests on m_1 objects with the property and m_2

objects without the property. Then the goal of prediction is to interpret the combined k test results to decide whether the object has or does not have the property of interest. The interpretation is in the form of a probability statement concerning the potential presence or absence of the property.

The CPBS approach is composed of a set of methodologies that we have found useful in addressing both battery selection and battery interpretation in the problem of cancer hazard identification. Although the CPBS was initially developed to address the problem of how to utilize short-term tests in screening for potential chemical carcinogens (see Section 1.3 of this chapter for a description of the cancer hazard identification problem), we have found that this collection of methods has a much broader potential for application. For example, the CPBS approach can be used for selecting and interpreting tests in such diverse areas as medical diagnosis, quality control, and detection of environmental hazards. Thus we have written this book in a "generic problem" context rather than in the cancer hazard problem context. We utilize the generic terms "objects," "tests," and "properties" rather than using the terms related to a specific problem. In order to provide examples of how the methods can be applied in practice, we have included specific examples derived from our experience with the cancer hazard identification problem.

1.2. OVERVIEW OF THE CPBS METHODOLOGY

Consider the following generic problem: a decision maker has an object for which he is trying to determine whether a particular property is present or absent. Suppose, further, that he has available to him a number of different tests, each with a different cost and degree of effectiveness in identifying the property of interest. The CPBS methodology can help the decision maker in two major ways: (1) it can assist him in formulating and selecting the best battery of tests to use and (2) it can provide a means for interpreting the results of a battery of tests. In order to accomplish these two goals, the CPBS requires knowledge about the effectiveness (performance) of each of the tests in identifying the property of interest. We assume that a data base is available containing test results on objects with known properties, so that performance measures for each test and combination of tests can be estimated. The CPBS then follows the scheme depicted in Fig. 1.2.1.

The CPBS begins with the analysis of any data base containing the results of tests on objects with known properties (Fig. 1.2.1A*). Both the

* Note: The references to Figs. 1.2.1A–1.2.1J are used to designate the portions of Fig. 1.2.1 that are marked with the letters A–J.

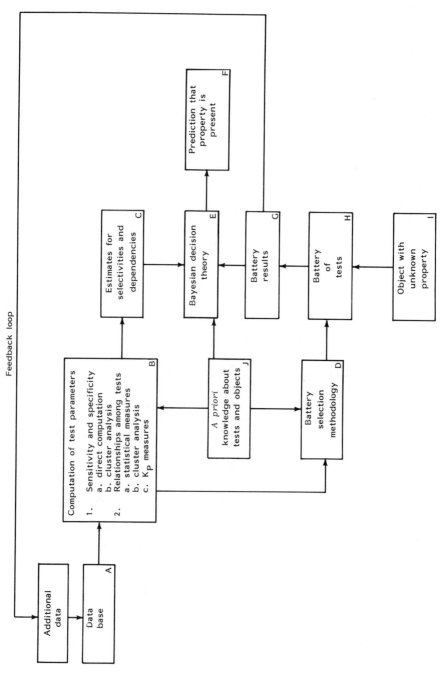

FIGURE 1.2.1. A schematic diagram of the CPBS approach.

selection of the best battery of tests and the prediction of the presence or absence of a property of interest based on the interpretation of the results of a battery of tests depend very much on the performance (reliability, reproducibility, and accuracy) of the individual tests. Thus we begin by determining appropriate performance measures for each individual test (Figs. 1.2.1B1). The performance indices that can be determined from most data bases are *sensitivity* (fraction of objects with the property of interest that were correctly identified by the test) and *specificity* (fraction of objects without the property of interest that were correctly identified by the test). These two indices together are indicative of the *selective* ability of the test (Fig. 1.2.1C), which is the ability of the test to detect both the presence and absence of the property of interest. If a data base is available that contains a sufficiently large number of results of each of the tests on both objects with and without the property of interest, then the sensitivity and specificity for each test can be easily estimated by direct computation from their definitions (Fig. 1.2.1B1a). In a less ideal data base possessing insufficient test results on objects with known properties (i.e., where reliable direct estimates of either or both of these indices are impossible), an exploratory data analysis tool called cluster analysis may be useful for this purpose (Fig. 1.2.1B1b). The accuracy of these direct and/or indirect estimation procedures may be further enhanced if *a priori* knowledge about the tests (i.e., knowledge about the selective ability of the tests that comes from other sources besides the data base) is available (Fig. 1.2.1J).

The interpretation of the results of a battery of tests not only depends on the individual performances of the tests, but can also depend on whether relationships (interdependencies) exist among the tests (Fig. 1.2.1B2). The presence of dependencies among tests can be detected by common measures of association such as the Chi-square statistic or the cross product ratio (Fig. 1.2.1B2a) if the data base has a sufficiently large number of results on objects both with and without the property of interest. When the data base is less ideal, then cluster analysis can be utilized to provide empirical evidence that relationships exist among the tests (Fig. 1.2.1B2b). If dependencies are found to exist among the tests in a battery, then these dependencies should be quantified so that the performance of the battery and the predictions based on battery results can be accurately computed. This is accomplished for each pair of tests through the estimation of measures of dependence (Fig. 1.2.1B2c). New dependency measures known as the K_p measures which are suitable for use in the methods used by the CPBS for battery selection and battery interpretation are described in this book (see Section 3.4).

At this point the CPBS has provided a means for estimating the sensitivity and specificity of each of the tests and the dependencies among the tests from information contained in a data base. We call this the *preliminary analysis* of the data base. (A detailed discussion of this process is given in Chapter 3.) After preliminary analysis has been completed, we are ready to address the generic problem stated at the beginning of this section. The estimates for the sensitivities, specificities, and dependencies provide the basis for solving the decision maker's problems. First, together with *a priori* information about the "costs" of the tests (e.g., dollar cost, difficulty, manpower requirements), they constitute the basis on which the decision maker will decide which battery of tests to use [e.g., one that is highly selective and predictive as well as cost effective (Fig. 1.2.1D)]. (A detailed description of how *battery selection* can be accomplished is given in Chapter 4.) Second, they provide a basis for predicting the presence or absence of a property of interest (Fig. 1.2.1E).

For the latter problem, after the decision maker has selected an appropriate battery to use, the object with unknown properties is tested by this battery (Fig. 1.2.1H). The results of this battery (Fig. 1.2.1G), along with the estimates of the selectivities and interdependencies among the tests in the battery (Fig. 1.2.1C) and an *a priori* probability that the object has the property of interest (Fig. 1.2.1J) are used to compute the prediction through the use of the well-known probability law called Bayes' theorem (Fig. 1.2.1F). The output of this process is a probability that the property is present in the tested object. (This *battery interpretation* process is described in Chapter 5.) The decision maker must then decide whether this probability is strong enough to warrant action or whether more testing may be required and the process repeated.

As additional information about the tests is gathered and entered into the data base, the estimates for the selectivities and dependencies improve (feedback loop from Fig. 1.2.1G to Fig. 1.2.1A). This in turn strengthens the reliability of the predictions and improves the overall decision-making process.

The preceding description of the CPBS approach was given in generic terms. The following section describes the cancer hazard identification problem and describes the CPBS approach in the context of this problem in order to further highlight the basic steps.

1.3. THE CANCER HAZARD IDENTIFICATION PROBLEM

More than two centuries ago, Percival Pott identified that exposure to tar and soot was related to the high incidence of scrotal cancer in chimney sweeps (Pott, 1775). Since that first observation of a chemical substance with the capacity to cause cancer, many other chemical carcinogens have been identified through analyses and occupational exposure studies of humans, together with experimental results on laboratory animals (U.S. Congress, 1981). It is estimated that there are roughly 60,000–70,000 commonly used chemicals (Shrader-Frechette, 1985; NAS, 1984) and that approximately 1,000 new chemicals are introduced each year (Shrader-Frechette, 1985). Yet we only have information about health effects for approximately 10%–15% of the chemicals (NAS, 1984). Thus the potential hazard of the vast majority of chemicals introduced into our drugs, foods, consumer goods, and environment remains unknown.

It is very important that chemical health hazards currently facing U.S. workers and the general public can be identified so that regulatory agencies can take appropriate measures to prevent or reduce human exposure to the higher-risk chemicals. It is also important for the producers of new chemical products to identify potential health hazards early on in the development of products, so that development of higher-risk chemicals can be halted or regulated before wasting time and money. Both types of actions would result in a reduction in the amounts of hazardous substances introduced into the environment and consequently would result in an overall reduction in human health risks.

Five methods are commonly used to identify mutagenic and carcinogenic hazards (Shrader-Frechette, 1985): (1) use of case clusters (i.e., noting an unusual concentration of cases of the same type and attempting to find the common cause), (2) use of epidemiological techniques (i.e., to show a positive association between agent and disease), (3) comparison of compounds in terms of structure–toxicology relationship (i.e., to compare an agent's chemical properties with known carcinogens in order to obtain evidence of potential carcinogenicity), (4) long-term animal bioassays (i.e., exposure studies on live animals, usually rodents of both sexes and two different animal species), and (5) short-term *in vitro* or *in vivo* tests (i.e., laboratory testing of bacteria, cultured mammalian cells, or whole animals in order to identify mutagenic, DNA damaging, and oncogenic effects). Methods (1) and (2) are generally applied after the fact—i.e., evidence is gathered after an unusually high incidence of a particular disease is noted. Methods (3) and (5) have been used to explore potential associations between agent and disease; and method (4) has generally been used to test the hypothesis that a particular agent is the cause of a particular disorder.

At the present, in the United States, regulatory decisions are based primarily on the results of animal carcinogenicity bioassays. Only about 1000 chemicals have been subjected to adequate rodent bioassays for carcinogenicity, in part because such testing is both time consuming and expensive. For every chemical tested, an animal bioassay requires two or more years to complete, and the cost approaches one million dollars

(Gold et al., 1984, 1986; Nesnow et al., 1987). Such figures clearly indicate that it would be infeasible to use animal bioassays to screen chemicals for carcinogenicity. This realization has led to the development of methods based on chemical properties [e.g., the computer automated structure evaluation (CASE) methodology (Klopman and Rosenkranz, 1984; Rosenkranz et al., 1984)] and methods based on short-term in vitro and in vivo tests, which constitute the basis for the CPBS approach.

Short-term tests that measure genotoxicity or some other end point related to the carcinogenic process have the potential to be very useful in predicting the carcinogenicity of chemicals. These tests are inexpensive (ranging from $1000 to $50,000) and rapid (one week to a few months). However, no short-term test has proven to be a perfect surrogate for the animal cancer bioassay, and the problem of selecting tests and interpreting their results has been the subject of considerable controversy (Ashby, 1986; Tennant et al., 1986; Haseman et al., 1986; Weisburger and Williams, 1986; Lave and Omenn, 1986).

Given the multistage, multicausal nature of the carcinogenic process, it is not surprising that no short-term test has been developed that corresponds exactly to carcinogenesis in rodents. To compensate for this problem, a combination or battery of tests is usually performed on a chemical with unknown carcinogenicity. Two issues result: what tests should be included in a battery, and how the results should be interpreted. The main portion of this book describes in detail the procedure we have developed to answer these questions—the carcinogenicity prediction and battery selection (CPBS) method (Chankong et al., 1985; Rosenkranz et al., 1984). In the remainder of this section, the CPBS will be described in the context of the cancer hazard identification problem. In addition, other approaches to this problem are outlined, for comparison with the CPBS and to illustrate the controversies that exist at present.

Consider the following scenario: A chemical company is in the early developmental stages of a new product, and the company manager must have information concerning the carcinogenic potential of the chemical so that he can decide whether to continue the development of the product and whether any special safeguards should be taken. At this stage of product development, a full-fledged animal bioassay is deemed to be too time consuming and costly, so an alternate approach using short-term in vitro and in vivo tests (assays) has been selected. Suppose that the company has a choice of several possible assays and that a data base containing the results of these assays on known carcinogens and known noncarcinogens is available. The problems facing the manager (and his analysts) are (i) which of the available tests should be used, and (ii) how should the results of a particular battery be interpreted?

The CPBS approach to answering these two questions begins by examining how well each of the short-term tests worked on the known carcinogens and noncarcinogens in the data base, thus providing an indication of the ability of each test to separate the carcinogens from the noncarcinogens. This is accomplished through direct computation of the sensitivity (proportion of carcinogens correctly identified) and specificity (proportion of noncarcinogens correctly identified) for each test in the data base, as well as through the use of cluster analysis. The joint responses of the tests on the chemicals in the data base are examined in order to evaluate how well various combinations of tests would work, and to see whether redundancies exist among the tests. This is accomplished through the use of various statistical measures that examine the dependencies that exist among the tests. At this point the CPBS assists the decision maker in choosing a preferred combination of tests by providing an analysis of the abilities of selected combinations in separating carcinogens from noncarcinogens. The CPBS also assists the decision maker in making trade-offs between testing cost and battery performance for various combinations of tests. After a particular battery of tests has been selected and applied to a chemical whose potential for carcinogenicity is not known, the CPBS computes a probability that the chemical is a carcinogen based on the test results, through the use of Bayes' theorem. This probability is given as a function of the prior probability (likelihood before testing) that the particular chemical is a carcinogen. After the decision maker has the prediction based on a specific battery's results, the CPBS can help him/her evaluate whether additional testing would be worth the additional time and effort.

Other approaches to selecting and utilizing batteries of short-term tests have been proposed. For example, Weisburger and Williams (1986) proposed a "decision point" approach, in which tests are performed in order of increasing phylogenetic complexity (e.g., bacterial, insect, and mammalian test systems) and end point (e.g., DNA damage, mutation, chromosomal effects, and transformation). At each stage, all the information available about the responses of the chemical is evaluated.

John Ashby (1986) has emphasized the importance of *in vivo* tests performed using whole living animals. His scheme also uses a tier of tests, starting with the *Salmonella* mutation or Ames assay (Sty). If Sty is negative, another *in vitro* test is performed, using a different activation system, target organism, and end point. If Sty and the second *in vitro* test are both negative, then, in Ashby's scheme, testing would stop and the chemical would be considered to pose a negligible threat. If Sty is positive or if the second *in vitro* test is positive, then the second tier of *in vivo* tests is performed. First, the effect of the chemical on the chromosomes in the bone marrow of mice would be tested, using a procedure called the micronucleus test (Mnt). If

Mnt is negative, then a second *in vivo* test would be performed, probably testing for induction of DNA repair in the liver of a mouse or rat. If both of the *in vivo* tests are negative, then Ashby feels that the potential of the chemical to cause genetic damage indicated by the *in vitro* test has been shown not to be realized in the whole animal, and that therefore the chemical should not be considered a potential carcinogen. If one of the *in vivo* tests is positive, then the chemical should be treated as a carcinogen. At present, few chemicals have been tested in the *in vivo* assays. Thus, it is difficult to evaluate whether *in vivo* tests do in fact have as critical a role in a testing program as Ashby suggests. However, the main points of Ashby's approach have been accepted as guidelines for mutagenicity testing by the Canadian Ministers of National Health and Welfare and of the Environment (ACM, 1986).

A committee of the International Commission for Protection against Environmental Mutagens and Carcinogens has been developing a procedure for evaluating test results from any given set of tests. Short-term tests are given weighting factors determined by relevance to risk in humans (Brusick *et al.*, 1986). Thus, a positive result in a test using mammalian cells, for example, would be given greater weight than a positive result in a test using bacteria. In addition, in general, negative results are given less weight than positive results. An overall score is computed for each chemical in this scheme to represent the integration of all of the testing information. Criteria for assessing the practicality of this scheme are being developed (Brusick *et al.*, 1986).

Another approach for displaying results from a series of tests is the production of "genetic spectra," in which each test result is converted into a potency, the lowest effective dose (LED) for positive results and the highest ineffective dose (HID) for negative results (Waters *et al.*, 1986). Patterns of responses can be detected by comparing the genetic spectra for a given set of chemicals. Genetic spectra have been generated as part of the evaluation of 250 chemicals by the International Agency for Research on Cancer.

Lave and Omenn (1986) have developed a procedure for evaluating batteries of short-term tests that has some similarity to the CPBS. The main difference is that test results are analyzed by logit analysis for correspondence to carcinogenicity and noncarcinogenicity. This analysis generates a score between zero and one, which represents a probability of carcinogenicity, similar to CPBS. Batteries are then evaluated on the basis of the trade-offs between the costs of the testing and the societal costs of false positives (chemicals that are not carcinogens but have been falsely labeled as carcinogens) and false negatives (chemicals that are carcinogens but have been falsely labeled as noncarcinogens). Lave and Omenn demonstrated that batteries of tests with high predictive capabilities (i.e., 80% or greater) can be cost effective.

In summary, the need for inexpensive and reliable procedures for testing the potential carcinogenicity of chemicals is apparent, because of the large number of chemicals in commerce and industry that have never been tested and because of the cost of animal bioassays. A number of schemes have been proposed for selecting and interpreting short-term test results; however, for the most part evaluation of the schemes has been performed using only a few chemicals. We feel that past data reflecting the ability of the tests to separate carcinogens from noncarcinogens should be used in selecting which battery of tests to use and in interpreting the results of the battery. The CPBS method was developed using a data base of over 2000 chemicals, and it has been tested numerous times using independent data sources. Although we contend that information about the performance of tests is of critical importance, this does not mean that the selected battery cannot or should not also satisfy biological criteria. In fact, the CPBS is compatible with some of the above-mentioned procedures for battery selection; for example, it is possible to design a battery that fulfills the biological criteria of Weisburger and Williams and that also satisfies the test performance guidelines developed through the CPBS method. This possibility, along with other examples of the CPBS method for cancer hazard identification, is presented throughout the book; and Chapter 6 is devoted completely to this topic.

REFERENCES

ACM (Advisory Committee on Mutagenesis), 1986, *Guidelines on the Use of Mutagenicity Tests in the Toxicological Evaluation of Chemicals*, Minister of National Health and Welfare and Minister of the Environment, Ottawa, Canada.

Ashby, J., 1986, "The prospects for a simplified and internationally harmonized approach to the detection of possible human carcinogens and mutagens," *Mutagenesis*, **1**:3-16.

Brusick, D., Ashby, J., de Serres, F., Lohman, P., Matsushima, T., Matter, B., Mendelsohn, M., and Waters, M., 1986, "Weight-of-evidence scheme for evaluation and interpretation of short-term results," In *Genetic Toxicology of Environmental Chemicals, Part B: Genetic Effects and Applied Mutagenesis*, C. Ramel, B. Lambert, and J. Magnusson (eds.), Alan R. Liss, New York, pp. 121-129.

Chankong, V., Haimes, Y. Y., Rosenkranz, H. S., and Pet-Edwards, J., 1985, "The carcinogenicity prediction and battery selection (CPBS) method: A Bayesian approach," *Mutation Res.*, **153**:135-166.

Gold, L. S., de Veciana, M., Backman, G. M., Magaw, R., Lopipero, P., Smith, M., Hooper, N. K., Havender, W. R., Bernstein, L., Peto, R., Pike, M. C., and Ames, B. N., 1984, "A carcinogenic potency database of the standardized results of animal bioassays," *Environ. Health Perspect.*, **58**:9-319.

Gold, L. S., de Veciana, M., Backman, G. M., Magaw, R., Lopipero, P., Smith, M., Blumenthal, M., Levinson, R., Bernstein, L., and Ames, B. N., 1986, "Chronological supplement to the Carcinogenic Potency Database: Standardized results of animal bioassays published through December 1982," *Environ. Health Perspect.*, **67**:161–200.

Haseman, J. K., Tharrington, E. C., Huff, J. E., and McConnel, E. E., 1986, "Comparison of site-specific and overall tumor incidence analyses for 81 recent National Toxicology Program carcinogenicity studies," *Regul. Toxicol. Pharmcol.*, **6**:155–170.

Klopman, G., and Rosenkranz, H. S., 1984, "Structural requirements for the mutagenicity of environmental nitroarenes," *Mutation Res.*, **126**:227–238.

Lave, L. B., and Omenn, G. S., 1986, "Cost-effectiveness of short-term tests for carcinogenicity," *Nature*, **324**:29–34.

NAS (National Academy of Sciences, U.S.), 1984, *Toxicity Testing: Strategies to Determine Needs and Priorities. The Report of Steering Committee on Identification of Toxic and Potentially Toxic Chemicals for Consideration by the National Toxicology Program, Board on Toxicology and Environmental Health Hazards, Commission on Life Sciences, National Research Council*, National Academy Press, Washington, D.C.

Nesnow, S., Argus, M., Bergman, H., Chu, K., Frith, C., Helmes, T., McGaughey, R., Ray, V., Slaga, T. J., Tennant, R., and Weisburger, E., 1987, "Chemical carcinogens: A review and analysis of the literature of selected chemicals and the establishment of the Gene-Tox Carcinogen Data Base," *Mutation Res.*, **185**:1–195.

Pott, P., 1775, "Chirurgical Observations," Reprinted in *Natl. Cancer Inst. Monog.*, 1963, **10**:7.

Rosenkranz, H. S., Klopman, G., Chankong, V., Pet-Edwards, J., and Haimes, Y. Y., 1984, "Prediction of environmental carcinogens: A strategy for the mid-1980's," *Environ. Mutagen.*, **6**:231–258.

Shrader-Frechette, K. S., 1985, *Risk Analysis and Scientific Method*, D. Reidel, Dordrecht.

Tennant, R. W., Stasiewicz, S., and Spalding, J. W., 1986, "Comparison of multiple parameters of rodent carcinogenicity and *in vitro* genetic toxicity," *Environ. Mutagen.*, **8**:205–227.

U.S. Congress, 1981, Office of Technology Assessment, *Assessment of Technologies for Determining Cancer Risks from the Environment*, U.S. Government Printing Office, Washington, D.C.

Waters, M. D., Stack, H. F., and Brady, A. L., 1986, "Analysis of the spectra of genetic activity in short-term tests," In *Genetic Toxicology of Environmental Chemicals, Part B: Genetic Effects and Applied Mutagenesis*, G. Ramel, B. Lambert, and J. Magnusson (eds.), Alan R. Liss, New York, pp. 99–109.

Weisburger, J. H., and Williams, G. M., 1986, "Rational decision points in carcinogenicity bioassays based on mechanisms of mutagenesis and carcinogenesis," in *Genetic Toxicology of Environmental Chemicals, Part B: Genetic Effects and Applied Mutagenesis*, C. Ramel, B. Lambert, and J. Magnusson (eds.), Alan R. Liss, New York, pp. 91–98.

Chapter 2

Fundamental Basics of the CPBS Approach

In this chapter we will examine four basic methodologies and decision tools that are utilized in the CPBS approach to decision making. The first is Bayesian decision analysis, which forms the heart of the CPBS approach. Tests and measurements that are used to identify or detect a property of interest are generally not perfect. When tests are biased or inaccurate, it is often advantageous to use more than one test. The interpretation of a combination of test results can be problematic because there often exists a variable amount of information overlap (positive dependence) and differences (negative dependence) among the tests. It is a difficult problem to account for both the imperfection of the individual tests as well as their interdependencies in their joint interpretation.

There are two general statistical approaches to making inferences (e.g., decisions of whether a property is present or absent) based on test results that are imprecise and/or inaccurate and interdependent. The first is to use classical statistics (e.g., hypothesis testing). Classical statistical techniques require reasonably accurate estimates of the joint probability distribution of test results. If such estimates can be made, then existing multivariate procedures, such as discriminant analysis, logit/probit analysis, and correlational analysis may be used. (For details about these statistical analyses, see, for example, Goldstein and Dillon, 1978; McKelvy and Zavoina, 1975; and Finney, 1971.) In many practical situations, the distributional assumptions required to perform these analyses (e.g., normality assumptions) are not satisfied and are thus too restrictive. Moreover, the required data may not be available and are often expensive and time consuming to obtain.

The second approach is to use Bayesian-based procedures. One characteristic of Bayesian-based procedures is that they allow both prior information (including subjective value judgment) and sampling information to be combined in the weighting scheme inherent in Bayes' formula. Thus, for example, in a clinical diagnosis problem a physician's expert judgment can be properly included with the test results in making the diagnosis. A second characteristic of Bayesian-based methods is that (under certain assumptions) they can be formulated in a recursive form. This means that Bayesian methods would allow successive updating of battery interpretation as additional test results are obtained, which would be particularly useful if sequential testing procedures are being considered.

We have chosen Bayesian decision analysis as the basis of the CPBS approach because (1) we feel that decision making can be improved by properly combining expert judgment and sampling information; (2) it is often desirable to make decisions in a sequential fashion so that testing costs can be reduced; and (3) (as will be shown in later chapters) the commonly used test performance measures are basic components in Bayes' formula. Section 2.1 of this chapter describes the basic formulation, concepts, and tools for performing Bayesian decision analysis.

A second basic methodology that is used in the CPBS approach is cluster analysis. Cluster analysis is an exploratory analysis tool (in contrast to hypothesis testing) that can be used to ascertain the fundamental differences and similarities among items in a data base. For example, cluster analysis can be used to determine which of the tests respond most similarly on the same set of objects, or alternatively, which of the objects are most alike based on the test results. This type of information can be used to determine, for example, (1) which of the tests should be used in a particular battery, (2) the number of basic types (groups) of tests, and (3) whether the objects can be separated (by their properties) on the basis of a particular battery of tests. A basic discussion of various cluster analysis techniques is given in Section 2.2 of this chapter.

A third element of the CPBS approach is multiple-objective decision making. The problem of choosing the "best" combination of tests to use for a particular problem involves many attributes of the tests, including (1) the ability of each test to identify a property of interest, (2) the costs of the tests, (3) the time it takes to complete each test, and (4) the amount of manpower and other resources required by each test. The importance of each of these attributes in the selection of the most appropriate battery of tests is dependent on the objectives of the person interested in using the tests as well as the constraints of the problem. For example, a pharmaceutical company developing a new product would most likely be most concerned about finding a battery of tests that performs well in detecting whether the product is safe for humans, whereas the cost of the battery might be of less importance. This same company may have limited time (because of competition from other companies) and resources (e.g., manpower and equipment) to do the testing.

In such circumstances there is generally no "best" battery (i.e., no battery of tests that is best in each of the attributes) and thus compromises must be made among the objectives. Multiple-objective optimization provides the decision maker with a means for finding the "best compromise" among the objectives. A general discussion of this approach is found in Section 2.3.

A fourth basic methodology found in the CPBS approach is dynamic programming. When a relatively large number of tests is available for detecting a property of interest and one is not sure of how many or which of the tests to use, the number of choices becomes very large. It requires an excessive amount of computation to list every possible combination of tests along with their associated attributes, and a more computationally efficient procedure would be useful. One way of reducing this computational problem is through the use of a sequential optimization technique called dynamic programming. Section 2.4 of this chapter provides a general discussion of this technique.

2.1. BAYESIAN DECISION ANALYSIS

Since this book is about decision making when multiple test results are available, a major part of the book is devoted to discussing how the test data should be organized and selected, and how they should be combined with other pieces of information to help the decision maker arrive at a decision. Many of the ideas are based on the view that an individual (or group of individuals) makes decisions by applying his/her *best* knowledge of the situation in combination with his/her preference. A decision is thus made based on the decision maker's state of knowledge (degree of belief) and state of mind (preference). Because beliefs and values of two different individuals are rarely the same, a decision that is considered good by one may not necessarily be so by the other. This view of decision making has a natural tie with the Bayesian approach, which prescribes how the degree of belief can be quantified and how new data can be combined with prior data to update the degree of belief. The aim of this section is to describe basic concepts and procedures of the Bayesian approach to decision making, which will in turn form a foundation to other parts of the book.

For brevity and clarity, the basic ideas will be presented by using the following example: A general manager of a chemical company is faced with the problem of having to decide whether to go ahead with the development and marketing of a new compound X proposed by his research and development division. If the compound proves to be noncarcinogenic its successful commercial development and marketing will bring substantial economic benefit in the order of $10 million per year. However, if the compound is later found to be carcinogenic there will be a great loss in terms of development costs, legal fees, and company reputation. The magnitude of loss will depend on how far along the development and marketing activities have progressed before the problem is discovered. We will assume the monetary equivalence of this loss to be in the order of $8 million per year.

This typifies a practical decision problem. It consists of the following elements: (1) a set of possible actions—Do Nothing (DONT) or Go Ahead with the development (GO); (2) a set of possible states of the world—chemical X is noncarcinogenic (NO) or chemical X is carcinogenic (CA); and (3) a set of outcomes, each of which represents the consequence of an action under a given state of the world. These elements constitute the basic structure of a decision problem. It is often convenient to display them in compact form, such as by the payoff matrix given in Fig. 2.1.1. Of course, if the "Do Nothing" action is taken, there will be no monetary gain or loss to the company, as reflected in the figure.

To make a choice, the manager will need to know two more factors: the likelihood that compound X is a carcinogen and an appropriate decision rule to use to compare alternatives. The former represents his degree of belief and the latter his preference and value. Different combinations of these two factors lead to different classes of decision problems, which are described below.

2.1.1. Making Decisions under Perfect or No Information

Two extreme situations are worth mentioning here. On one extreme, the manager and his research staff may have extensive knowledge about the compound so as to confidently declare it carcinogenic or not. In this situation of (complete) certainty, the manager may simply use the net monetary gain as the basis for making the decision. For example, if evidence overwhelmingly indicates that the chemical is not a carcinogen, the manager would quite likely choose the "Go Ahead" option, since it has more potential monetary gain than the "Do Nothing" option.

On the other extreme, nothing may be known at all about the chemical (although this is an unlikely situation). To make a decision under such complete uncertainty, the manager will choose a decision rule that best

	State of nature	
	Noncarcinogenic (NO)	Carcinogenic (CA)
Do Nothing (DONT)	0	0
Go Ahead (GO)	10	−8

FIGURE 2.1.1. Payoff matrix for the example (payoff in millions of dollars).

reflects his attitude toward risk. If he is a risk seeker, he will act just like most optimists or gamblers would, taking an action that could potentially yield the most benefit regardless of risk. In the above example, application of this optimistic decision rule would result in the selection of the "Go Ahead" option. A risk avoider, on the other hand would take a more pessimistic view of the world, and would thus follow a more conservative decision rule. Here, an action that gives the best return under the worst scenario will be selected. This is commonly known as the Pessimistic or Walds rule. Under the "Go Ahead" option, the worst that can arise is that the compound turns out to be a carcinogen and the company loses $8 million per year. Under the "Do Nothing" option, the worst net gain is zero. A risk-avoiding manager would thus prefer to do nothing.

The optimistic and pessimistic rules represent two extreme views of the world when no information is available. Most people, however, take intermediate positions and make decisions accordingly. The in-betweenist rule known as Hurwitz's rule and the Bayes–Laplace rule based on the principle of insufficient reason are examples of decision criteria that can be used for this purpose.

2.1.2. Making Decisions under Imperfect Information

The above two extreme states of knowledge of the world are often not realistic. More commonly, some partial information would be available so that an assessment of the likelihood of each state of the world would be possible. At this stage, important steps to be carried out using the Bayesian approach include (1) making an initial estimate of the likelihood of each state of the world based on prior information, (2) updating prior estimates using new information if available, (3) assessing the decision maker's preference and choosing an appropriate decision rule to combine beliefs and values, and (4) making the necessary analysis and comparative evaluation of alternatives. In the remainder of this section, important issues and basic procedures for implementing these steps are described and illustrated.

2.1.3. Subjective Probability and Prior Estimate

It is convenient to treat the likelihood that the chemical is a carcinogen as a probability. The manager's estimate of this probability will in turn be considered as a subjective probability reflecting his current state of knowledge. Classical statisticians usually have difficulty accepting the use of subjective probability, citing mathematical and philosophical reasons.

One of the main concerns is whether or not subjective probability satisfies normal probability laws that are so crucial for its successful applications. Fortunately, Savage's work (1954) provides a satisfactory resolution to this issue. It is shown that under some mild assumptions about an individual's decision-making behavior, both subjective probability and utility exist. They represent, respectively, the individual's degree of belief and preference, and together they can be used to model the individual's decision-making process based on the maximum expected utility principle. Such subjective probability indeed satisfies all probability laws.

Another issue of concern involves nonuniqueness. If subjective probability represents an individual's degree of belief, an event may have different probabilities depending on who assesses it. To workers in precision sciences, this may seem conceptually absurd and scientifically unacceptable. Nevertheless, subjective probability is increasingly accepted as having a useful place in decision making. It is perfectly acceptable for two individuals to make different decisions about an identical choice because they happen to have different beliefs and values. Yet both can be quite satisfied with their decisions. Moreover, in most practical cases, subjective probability may be the only appropriate concept to use. While both "objective" probability (either classical or frequency view) and subjective probability provide a numerical measure of uncertainty associated with an event, the "objective" concept is restricted to describing uncertainty caused by a natural chance process. Subjective probability, on the other hand, can easily handle uncertainty due to lack or imperfection of knowledge about the event, as often found in most real decision problems. For example, a pediatrician examining a one-year-old child with high fever in an emergency room will prescribe a treatment based on his/her estimate of the probability that the child has a particular disease, say bacteremia. At the time the decision must be made, the child either has or does not have bacteremia. There is nothing random about the event. Yet the physician is still faced with uncertainty, due mainly to his/her inability to diagnose precisely from the information available. Obviously, if the physician could wait for two days for the result of a blood culture test, perfect knowledge of whether the child has bacteremia would be ascertained and all uncertainty removed. The physician's initial assessment of the probability that the child has bacteremia must therefore be based on the subjective probability concept.

Still another example illustrates the appropriateness of subjective probability. Suppose a ball is drawn from a box known to contain 7 black balls and 3 white balls. You are asked to guess the color of the drawn ball. A correct guess will yield a $1 prize, whereas an incorrect guess will cost you $1. Which color would you guess? Very likely, as with most other rational people, you would guess black, because the ball would appear to have more chance of being black (0.70) than white (0.30). What you have just used is the subjective probability concept. Indeed, if you insist on using objective probability, you would be going nowhere. Since the ball has already been drawn and is no longer subject to a chance process, the objective probability that the ball is black is either zero or one. Had the game been played before the ball was drawn, either the classical or frequency concept of probability would have been applicable, and the 0.70 probability of a black ball being drawn could indeed be used as a basis for making your decision. For the subjective probability school of thought, there would be no distinction between "before" and "after" the event. The act of drawing the ball does not add or delete any pertinent information available to you as a player. The subjective probability estimate of the drawn ball being black therefore remains at 0.70.

By the same reasoning, the interpretation and assessment of the probability that compound X in our example is a carcinogen is best considered from the subjective probability viewpoint. The classical view of probability, which relies on some "symmetry" argument, does not appear applicable. To interpret such probability using the relative frequency view is conceptually possible, but it would be an incorrect interpretation. For example, the statement "the probability that chemical X is a carcinogen is Pr(CA)" might *incorrectly* be interpreted as follows: when a large number of individuals are exposed to a toxic dose of chemical X under similar circumstances, 100Pr(CA)% of those individuals will develop cancer. Though this might appear to be a reasonable interpretation, the probability cannot be translated into the number of people at risk, since this risk would be dependent upon dose, duration, mode of administration, homeostatic mechanisms, and many other factors! Consequently, the only appropriate interpretation for Pr(CA) would be as a subjective probability that would indicate the uncertainty or imperfection in our state of knowledge concerning the carcinogenicity of chemical X.

The above discussion should help motivate and justify the use of subjective probability in decision making. This in turn supports the use of the Bayesian approach, which provides an appropriate mechanism and interpretation for combining new and current data to update the state of knowledge. This will be discussed next. For further reading on the appropriateness of the Bayesian approach and related philosophical discussions, see Good (1978), Hamaker (1977), Moore (1978), and Savage (1962). Also, for practical procedures for assessing subjective probability, see, for example, Winkler (1972), Berger (1985), and Buchanan (1982).

2.1.4. Bayesian Update

Suppose the manager has made an initial estimate of Pr(CA) to be 0.60 based on current information. Suppose further that a new test, A, for testing carcinogenicity has just been developed. The new test is capable of correctly identifying a carcinogen 80% of the time and a noncarcinogen 90% of the time. Suppose it yields a positive result when used on compound X. How should Pr(CA) be revised? Bayes' theorem provides a convenient device for this revision if the measures of test reliability specified above are given proper probabilistic interpretations. More specifically, we define the *sensitivity* of a test as the probability that the test will show a positive result if used on a carcinogen, Pr(+|CA), and the *specificity* as the probability that the test will show a negative result when used on a noncarcinogen, Pr(−|NO). The sensitivity and specificity of test A above are thus 0.80 and 0.90, respectively. The revised probability we are seeking is the conditional probability that X is a carcinogen given that test A has indicated a positive result, Pr(CA|A = +). This is also known as the posterior probability (posterior to performing the test), whereas Pr(CA) is usually called the prior probability, for an obvious reason. According to Bayes' theorem, we have

$$\Pr(CA|A = +) = \frac{\Pr(A = +|CA)\Pr(CA)}{\Pr(A = +)}$$

where Pr(A = +) is the marginal probability that test A will yield a positive result when used on X. Clearly, since X can either be a carcinogen, CA, or noncarcinogen, NO, we have

$$\Pr(A = +) = \Pr(A = +|CA)\Pr(CA) + \Pr(A = +|NO)\Pr(NO)$$

Noting that Pr(A = +|CA) is the sensitivity of test A, Pr(A = +|NO) is one minus the specificity of test A, and Pr(NO) = 1 − Pr(CA), we have

$$\Pr(A = +) = \Pr(A = +|CA)\Pr(CA)$$
$$+ [1 - \Pr(A = -|NO)][1 - \Pr(CA)] \qquad (2.1.1)$$

Fundamental Basics of the CPBS Approach

and

$$\Pr(CA|A = +)$$

$$= \frac{\Pr(A = +|CA)\Pr(CA)}{\Pr(A = +|CA)\Pr(CA) + [1 - \Pr(A = -|NO)][1 - \Pr(CA)]} \quad (2.1.2)$$

Since all quantities on the right-hand side of Eq. (2.1.2) are known, the revised probability can be computed. Using the given numerical values in the example, $\Pr(CA|A = +)$ is estimated to be

$$\Pr(CA|A = +) = \frac{(0.80)(0.60)}{(0.80)(0.60) + (1 - 0.90)(1 - 0.60)} = 0.92$$

That is, after combining prior information with the new test result, the manager now believes that the chance of X being a carcinogen is 0.92 as opposed to 0.60 prior to the test. The positive result of test A has reduced the degree of uncertainty, as viewed by the manager, by approximately 54%. The manager may now feel comfortable enough to make a decision or he may feel that further testing should be performed to reduce the degree of uncertainty further.

To illustrate how additional test results can be incorporated, suppose a new test, B, is used and the result is also positive. If the sensitivity and specificity of test B are 0.85 and 0.75, respectively, what should now be the revised probability that X is a carcinogen? That is, what is the perceived value of $\Pr(CA|A = +, B = +)$? Following the Bayesian updating strategy discussed above, the result of test B yields new information to be used to update the current estimate $\Pr(CA|A = +)$. The updating formula (2.1.2) can thus be used, with the original prior $\Pr(CA)$ replaced by the current prior $\Pr(CA|A = +)$ and the previous posterior $\Pr(CA|A = +)$ replaced by the new posterior $\Pr(CA|A = +, B = +)$. This gives

$$\Pr(CA|A = +, B = +)$$

$$= \frac{\Pr(B = +|[CA|A = +])\Pr(CA|A = +)}{\Pr(B = +|[CA|A = +])\Pr(CA|A = +) + [1 - \Pr(B = -|[NO|A = +])][1 - \Pr(CA|A = +)]}$$

If we further assume that tests A and B give results statistically independently, then $\Pr(B = +|[CA|A = +])$ is equal to the sensitivity of B, $\Pr(B = +|CA)$, and $\Pr(B = -|[NO|A = +])$ is equal to the specificity of B, $\Pr(B = -|NO)$. This simplifies the above updating formula to be

$\Pr(CA | A = +, B = +)$

$$= \frac{\Pr(B = +|CA)\Pr(CA|A = +)}{\Pr(B = +|CA)\Pr(CA|A = +) + [1 - \Pr(B = -|NO)][1 - \Pr(CA|A = +)]} \quad (2.1.3)$$

After known values of the right-hand side quantities are substituted, the revised probability after two tests becomes

$$\Pr(CA | A = +, B = +) = \frac{(0.85)(0.92)}{(0.85)(0.92) + (1 - 0.75)(1 - 0.92)} = 0.98$$

Again, the positive result of test B has strengthened the likelihood that X is a carcinogen. Naturally, more tests can be performed and results utilized in the recursive fashion shown above. This idea is also used with more general formulas in Chapter 5 of this book.

Instead of a sequential scheme, an alternative "batch" updating scheme can also be used. This is done by viewing the two tests as a combined source of information which can yield four possible outcomes: $(A = +, B = +)$, $(A = +, B = -)$, $(A = -, B = +)$, and $(A = -, B = -)$. This information, together with the assumed independence, produces the following:

$\Pr(CA | A = +, B = +)$

$$= \frac{\Pr(A = +, B = +|CA)\Pr(CA)}{\Pr(A = +, B = +|CA)\Pr(CA) + \Pr(A = +, B = +|NO)\Pr(NO)}$$

$$= \frac{\Pr(A = +|CA)\Pr(B = +|CA)\Pr(CA)}{\Pr(A = +|CA)\Pr(B = +|CA)\Pr(CA) + [1 - \Pr(A = -|NO)][1 - \Pr(B = -|NO)][1 - \Pr(CA)]}$$

Again with the numerical values given, we have

$$\Pr(CA | A = +, B = +) = \frac{(0.80)(0.85)(0.60)}{(0.80)(0.85)(0.60) + (1 - 0.90)(1 - 0.75)(1 - 0.60)}$$

$$= 0.98$$

which provides the same result as before.

If one or both tests have negative results, the updating formulas will need to be modified in a simple way. We will leave this to the reader. See also Chapters 5 and 6 for some examples of this situation.

Fundamental Basics of the CPBS Approach

The above discussion should further illustrate how the Bayesian approach emphasizes and uses two major sources of information as the basis for decision making. One is from sampling observation (e.g., test results) and the other is from our general (prior) knowledge about the event (compound X in our example), which may be generated through some other considerations. The Bayesian approach gives proper weights to these two types of information depending on their relative strengths, reliabilities, and accuracies. This is in contrast to classical statistics, where only sampling results are used with no due regard given to information that may come from other sources. To deny the use of prior information, if it exists, is to create what cognitive psychologists would call "representative bias"—putting too much emphasis on a specific piece of information (i.e., sampling information) and disregarding other more general information. In the Bayesian approach, both subjective and objective evidence can be properly utilized to upgrade our knowledge to a level where a *sound* decision can be made. If a certain threshold of knowledge level is set and a decision is made only if the threshold is reached or surpassed, the effect of the initial choice of prior probability on the final decision should be minimal. Indeed, the only real impact of such a choice would be on the amount of "new" information required to upgrade our knowledge to pass that threshold.

2.1.5. Choosing a Decision Rule and Making the Analysis

Having obtained the prior or revised probability values of the states of the world, the next step is to choose an appropriate decision rule followed by an analysis and evaluation. The most common decision rule is the "maximum expected utility" rule, whose mathematical foundation and motivation were developed by Von Neumann and Morgenstern (1953) and Savage (1954). The approach usually involves an assessment of the decision maker's utility function as a quantitative measure of his/her preference. The expected utility for each action is then computed and the action with highest expected utility value selected. For example, if $u(\$0m)$, $u(\$10m)$, and $u(-\$8m)$ are the manager's utility values measured in some utile scale of the three possible outcomes $0m, $10m, and -$8m, respectively, the expected utility values of the two options using only the prior information are

$$E(DONT) = \Pr(CA)u(\$0m) + \Pr(NO)u(\$0m)$$

$$E(GO) = \Pr(CA)u(-\$8m) + \Pr(NO)u(\$10m)$$

or, when test A is used,

$$E(\text{DONT}) = \Pr(\text{CA}|A = +)u(\$0\text{m}) + \Pr(\text{NO}|A = +)u(\$0\text{m})$$

$$E(\text{GO}) = \Pr(\text{CA}|A = +)u(-\$8\text{m}) + \Pr(\text{NO}|A = +)u(\$10\text{m})$$

The Go Ahead option would be recommended if $E(\text{GO}) > E(\text{DONT})$; otherwise the Do Nothing option would be more appealing. Assessing the utility function for use here is important, but it is often a tricky task. Knowing the typical relationship between the general shape of the utility function and the decision maker's attitude toward risk, such as the example in Fig. 2.1.2, is often helpful. For a detailed procedure for assessing the utility function, see Keeney and Raiffa (1976) and Chankong and Haimes (1983).

If the outcome is measured in monetary terms, as in our example, and if the decision maker is risk-neutral, meaning that his/her preference is directly proportional to the monetary outcome (see the middle graph in Fig. 2.1.2), then the monetary value itself can be used as a surrogate measure of the decision maker's preference. The corresponding decision rule is called the "maximum expected monetary value (EMV)" rule.

Using EMV in our example, we have:

with prior information only

$$\text{EMV}(\text{DONT}) = (0.60)(\$0\text{m}) + (0.40)(\$0\text{m}) = \$0\text{m}$$

$$\text{EMV}(\text{GO}) = (0.60)(-8\text{m}) + (0.40)(\$10\text{m}) = -\$0.8\text{m}$$

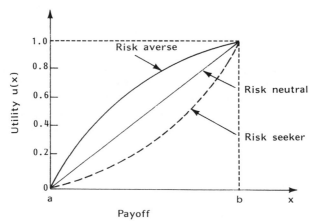

FIGURE 2.1.2. Risk behavior and utility functions.

Fundamental Basics of the CPBS Approach

or with the result of Test A incorporated

$$EMV(DONT) = (0.92)(\$0m) + (0.08)(\$0m) = \$0m$$

$$EMV(GO) = (0.92)(-8m) + (0.08)(\$10m) = -\$6.56m$$

Thus in both cases the Do Nothing option is recommended, since EMV(GO) < EMV(DONT).

2.1.6. Decision Tree and Worth of Information

Finally, we illustrate how a graphic device known as a decision tree can be used to help organize the data and guide the analysis in a more complicated situation.

Suppose test A has to be performed at some cost. Relevant questions to ask are: If test A is used, what strategy should be followed; that is what action should be taken for a given result of A? Should test A be used at all (if its cost is known)? What would be the maximum worth of test A? And finally, if test A is known to be a gold standard always yielding perfect information about the state of nature, what would be its maximum worth? The first question has been partially answered in the previous discussion. To answer the question completely, the best action when test A gives a negative result needs to be determined through similar computational steps. The second and third questions deal with the worth of test A, whereas the fourth question asks about the worth of perfect information. Dealing with these and other related questions usually involves multistage or sequential decision making and will benefit from using a decision tree.

A decision tree consists of decision nodes denoted by square boxes, outcome nodes denoted by small circles, action branches emanating from decision nodes, outcome branches emanating from outcome nodes, and terminal branches, which are a special form of outcome branches always ending with values of payoff or utility. A decision tree begins with a decision node and ends with terminal branches. A column of decision nodes signifies the beginning of the next stage of the decision. Figure 2.1.3 depicts a two-stage decision tree typifying the problem considered in this example. The first stage involves the decision on whether test A should be used, as reflected by the two decision branches. This is often called a preposterior analysis. The second stage involves determination of the best action strategy once it has been decided that the test should be used. This is known as a terminal or posterior analysis.

To draw a decision tree, we must first have clearly in mind the order in which the decisions will actually be made in practice. A decision tree is

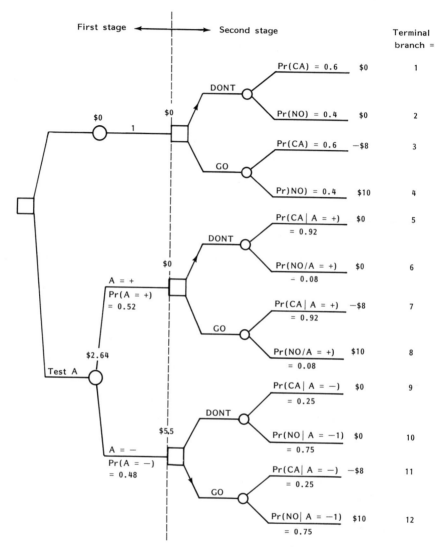

FIGURE 2.1.3. Decision tree used for the example. (Dollar amounts are in millions.)

then drawn with stages arranged to reflect this natural sequence of decisions. In our example, the manager has to decide first whether to commission the use of test A, followed by a decision on what to do with various test results once the test is used, hence the tree depicted in Fig. 2.1.3. The actual drawing begins with the initial decision node with as many decision branches emanating from it as there are alternative actions—to use or not to use

Fundamental Basics of the CPBS Approach

test A. At the end of each branch, we attach an outcome node. From each of these outcome nodes, we draw as many outcome branches as there are possible states of nature. A decision node is then assigned to each outcome branch, to begin the next stage. The process is then repeated at each decision node until the final stage. Finally, at the tip of each terminal branch, the corresponding payoff or utility is assigned.

The next step is to compute the probability associated with each outcome branch, which represents the probability of getting that particular outcome conditional on the actions and outcomes on the branches leading to it. For example, the probability associated with terminal branch 1 is the probability that X is a carcinogen given that test A is not used, which is simply the prior probability $\Pr(CA)$. The same is true for terminal branch 3, since actions DONT and GO do not affect this probability. For terminal branch 5, the associated probability is the probability that X is a carcinogen given that test A has been used and yielded a positive result. This is $\Pr(CA|A = +)$, which was computed earlier to be 0.92. The same probability is also assigned to terminal branch 7, since again actions DONT and GO do not affect this probability. For terminal branches 9 and 11, the required probability is

$$\Pr(CA|A = -) = \frac{\Pr(A = -|CA)\Pr(CA)}{\Pr(A = -|CA)\Pr(CA) + \Pr(A = -|NO)\Pr(NO)}$$

$$= \frac{(1 - 0.80)(0.60)}{(1 - 0.80)(0.60) + (0.90)(1 - 0.60)} = 0.25$$

and for branches 10 and 12, $\Pr(NO|A = -) = 1 - \Pr(CA|A = -) = 0.75$.

Note that test A, being quite reliable, has a major impact on the decision maker's degree of belief. Learning that test A has yielded a negative result, the manager sees that the risk of X being a carcinogen has been reduced from 0.60 to 0.25.

Finally, for the outcome branch "$A = +$" in the first stage, we need the value of $\Pr(A = +)$, the marginal probability that test A, if used, will yield a positive result. As computed earlier by Eq. (2.1.1), this value is 0.52.

At this stage, the average-out-and-fold-back procedure is applied. This involves the following steps: First, compute the expected payoff or utility for each outcome node in the final stage. For example, the expected payoff associated with the last outcome node of the second stage is $(0.25)(-\$8m) + (0.75)(\$10m) = \$5.5m$. Next, at each decision node preceding those outcome nodes, choose a decision branch leading to the outcome node with the best expected payoff. For example, for the last decision node of the second stage, the GO action yields higher expected payoff ($5.5m) than the DONT option ($0m), and is thus selected. The corresponding best possible expected payoff associated with this decision node is therefore $5.5m. The process is repeated sequentially until the initial stage.

Once completed, the decision tree now contains the necessary information to answer all but one of the questions posed earlier. (The exception is the one on the value of perfect information, which will be discussed shortly.) For example, it is clear by comparing the two outcome nodes in the first stage that the expected payoff could be increased from $0m when test A is not used to $2.64m when test A is used. Thus if test A costs less than $2.64m to execute, it would be worth using. The maximum worth of test A, which is defined as the difference between the expected payoffs when using test A and when using no test, is thus $2.64m.

If test A is not used, the DONT action is recommended, since its expected payoff is higher than the GO action, and likewise for the case when test A is used and a positive result is obtained. When test A is used and shows a negative result, the GO action has a superior expected payoff, as noted before.

Finally, to compute the value of perfect information, we observe that if a perfect test is available, an action yielding the highest payoff or utility under the state of nature indicated (with certainty) by the result of such a test should be selected. For example, if a perfect test indicates that X is a carcinogen, we would select the DONT option, having no doubt that X is indeed a carcinogen. The probability of obtaining a payoff that is best under a given state of nature S is therefore equal to the probability that the perfect test indicates S. This in turn equals the prior probability of S itself. Using a perfect test in our example, the payoff will be $0m (best payoff under the CA state) with probability 0.60 and $10m (best payoff under the NO state) with probability 0.40. The expected payoff under a perfect test is thus $(0.60)(\$0m) + (0.40)(\$10m) = \$4m$. Compared to the $0m expected payoff under no information, the value of perfect information for this example is $4m − $0m = $4m.

In general, the value of perfect information is the difference between the expected payoffs when using perfect information and when using no information. The expected payoff under perfect information is computed as the sum (over all states of nature) of the products of the best possible payoff under each state and the prior probability of that state.

2.1.7. Closing Remarks

The above description of Bayesian decision analysis is only meant to be introductory, highlighting important philosophical issues and illustrating

Fundamental Basics of the CPBS Approach

basic computational steps. There are many good texts on the subject. For further details and applications, the reader may consult Raiffa (1968), Holladay (1979), Weinstein *et al.* (1980), Lindley (1985), Buchanan (1982), and Chankong and Haimes (1983).

2.2. CLUSTER ANALYSIS

A major portion of the CPBS approach involves the analysis of the results of tests on sets of objects with known properties. In addition to providing information about the past performances of the tests, such a database can also provide indications of structural relationships between the elements (i.e., test and objects) in the database. Cluster analysis is a tool for exploratory data analysis. It is designed to uncover natural groupings in data and initiate new ideas and hypotheses about phenomena being studied (i.e., it is a means for generating hypotheses rather than testing hypotheses statistically). It is a means whereby large masses of data can be analyzed with the ultimate goals of explaining, understanding, ordering, classifying, and structuring the data.

A wide variety of cluster analysis techniques are available. Each method is designed to search for an "optimal partition" of the data based on some criterion function (e.g., Euclidean distance, norms). The methods differ in their strategies for partitioning the data and the criterion used for "testing the optimality" of a partitioning of the data. For example, some methods begin with all data items in one large group and then subdivide the groups into a prespecified number of subgroups according to a prespecified criterion function; other methods begin with all data items in separate groups and then combine groups according to some prespecified criterion.

The cluster analysis methods are embedded in an overall exploratory data analysis methodology (see Fig. 2.2.1). The process of exploratory data analysis begins with a data base containing a set of measurements (e.g., test or sampling results) on a set of objects or data items (e.g., chemicals, patients, manufactured products). The data base is analyzed by applying the formal cluster analysis methodology, which consists of (1) data representation, (2) selection and application of cluster analysis method(s), and (3) cluster verification. The result from this analysis is a "picture of the structure of the data base." If more than one clustering method is applied, then each method groups or "clusters" the data items (tests or objects) according to a particular measure of "alikeness." Thus each method provides a different view of the concordance and discordance among the data items. These views of the structure of the data in a data base can aid in identifying potential relationships that appear to exist among the data items.

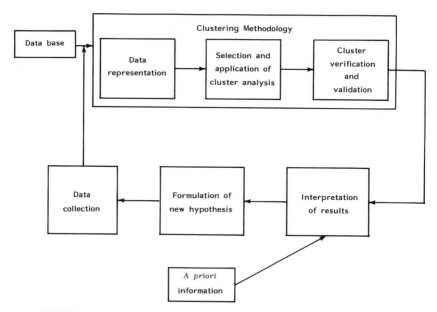

FIGURE 2.2.1. Methodological approach used in exploratory data analysis.

After completing the cluster analysis methodology, the next step is to explore the results and to try to explain or justify why the particular data items formed groups. In this step, information other than that found in the data base is considered. This additional information is used in formulating a hypothesis that is consistent with the cluster analysis results.

For example, suppose that the measurements are clinical tests on a set of patients and that we are interested in seeing which of the patients form groups based on the results of the tests. Suppose that after applying cluster analysis we find that essentially two groups of patients were formed. Further examination of the groups shows that one group consists of mostly sick patients (with a particular disease) and the second group consists of mostly healthy subjects. Then one might hypothesize that the clinical tests can be used to separate healthy patients from the sick patients. The results from cluster analysis *do not*, however, *indicate how* the tests can be used for this purpose or prove that the hypothesis is true.

As a second example, consider the cancer hazard identification problem and suppose that we are interested in exploring the relationships that exist among the short-term tests. [Note: in this example we are grouping the measuring devices (i.e., short-term tests) based on the results of the tests on a set of objects (i.e., chemicals), whereas, in the preceding example we were grouping the objects (i.e., patients) based on the results of the measuring devices (i.e., clinical tests). Cluster analysis can be used in either fashion, depending on whether we are interested in exploring relationships among the objects (e.g., patients or chemicals) or interested in exploring the relationships among the measuring devices (e.g., clinical tests or short-term assays).] Suppose that after applying the cluster analysis methodology on the short-term tests, we obtain several groups of tests. A closer examination of the groups indicates that the tests in each group are of the "same type" (e.g., one group consists mostly of bacterial assays, a second group consists mostly of chromosomal aberration assays). One might hypothesize that tests of the same type are positively dependent. Cluster analysis does not, however, indicate the strength of these dependencies nor does it prove that the hypothesis is true.

In both of the examples, we can see that the results of cluster analysis are qualitative and not quantitative, and that the validity of the results must be tested further. In general, one would collect more data and either repeat the exploratory data analysis to see whether similar results are obtained or apply statistical hypothesis-testing techniques to formally check the validity of the hypothesis. These additional tests can result either in the validation of the hypothesis or in the generation of new hypotheses; and in the latter case we would repeat the entire process over again.

Our objective in this section of the book is to provide a brief introduction to the cluster analysis methodology and the specific methods that we have employed in the CPBS. We provide a discussion of each of the three elements of the cluster analysis methodology: namely, data representation, selection and application of cluster analysis methods, and cluster verification/validation; and we illustrate some of the basic computational steps. There are several comprehensive books on the topic. For more details, the reader is referred to Anderberg (1973), Clifford and Stephenson (1975), Meisel (1972), and Dubes and Jain (1979).

2.2.1. Data Representation

As mentioned earlier, the raw data that exploratory analysis begins with consist of a set of measurements on a set of objects. This "matrix" of measurements on objects is called a *pattern matrix*, the term reflecting that each object can be represented as a point in an n-dimensional space according to the measurements ("n" equals the number of measurements). For example, if we have two clinical tests applied on a set of patients, one could graphically view the similarities among the patients by constructing a two-dimensional graph, using the range of values for the two test results as the two axes of the graph and then plotting the patients as points in the two-dimensional space. One could then view the patients as a "pattern" of points in this two-dimensional space and could visually see which of the patients group together (see, for example, Fig. 2.2.2). When more than two measurements are used, then we can no longer use this informal visual approach; and a more formal clustering algorithm is used.

The measurements in the database can be of a variety of measurement scales. For example, clinical test results can be expressed as either (1) continuous valued (e.g., blood pressure, oral or rectal temperature, white blood cell count) or (2) discrete valued [e.g., positive or negative results of a blood culture; positive or negative result for a particular short-term test applied to a chemical; having strong (=3), moderate (=2), weak (=1), or no (=0) response to a test].

In the case where the original measurements are continuous valued, a variety of cluster analysis techniques, called partitional clustering methods, can work with the pattern matrix directly. These methods divide or "partition" the objects into natural groupings based on the similarities among their "patterns" of measurements. Generally these methods require that the measurements for each test are first normalized so that each test result falls in the same numerical range.

When some or all of the measurements are discrete valued, the partitional methods cannot be used directly. In this case, the pattern matrix (raw data) is transformed to a so-called proximity matrix that contains continuous-valued elements; and then a class of clustering methods, called "hierarchical clustering methods," can be applied to the proximity matrix.

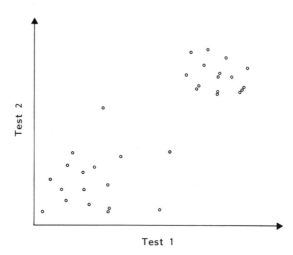

FIGURE 2.2.2. Graphical view of hypothetical patient data. The figure illustrates how each patient can be represented as a point in two dimensions based on the results of two tests.

Fundamental Basics of the CPBS Approach

In order to construct a proximity matrix from a pattern matrix, an index of proximity (e.g., similarity or dissimilarity) is established between each pair of data items. There are a variety of choices to use for an index of proximity, such as the proportion of matching responses between pairs of data items, sample correlation, or the Euclidean distance between pairs of data items. (For more examples, see Anderberg, 1973.)

As an example of this transformation of a pattern matrix to a proximity matrix, consider the hypothetical database or "pattern matrix" containing the discrete-valued results of 10 tests on 20 objects given in Table 2.2.1. To be more concrete, let us say that we are dealing with the cancer hazard identification problem, and thus the tests are short-term assays and the objects are chemicals. If we were applying cluster analysis to see which of the tests were most similar in terms of their responses on the chemicals, we could construct a proximity matrix by computing the similarity or dissimilarity between the responses of each pair of tests. For example, the similarity between two tests could be computed as the number of matching results between the two tests divided by the number of chemicals tested—the so-called "matching coefficient." Using this similarity measure on the data

TABLE 2.2.1. Hypothetical Data of Ten Tests on Twenty Chemicals

Chemical	T_1	T_2	T_3	T_4	T_5	T_6	T_7	T_8	T_9	T_{10}
1	+	+	−	+	−	−	+	+	+	+
2	+	+	+	+	+	−	+	+	−	−
3	+	−	−	+	+	−	+	+	−	+
4	−	+	+	−	+	−	+	−	−	+
5	+	+	+	+	+	+	+	+	+	+
6	−	−	−	−	−	−	−	−	−	−
7	−	+	+	+	−	+	+	+	−	+
8	+	+	+	+	+	−	+	+	+	−
9	−	+	+	−	+	+	−	+	+	−
10	+	+	+	+	+	+	+	+	+	+
11	−	−	−	−	−	−	−	−	−	−
12	−	+	+	−	+	−	+	+	−	−
13	+	−	+	+	−	−	+	+	−	−
14	+	+	+	+	+	+	+	+	+	+
15	+	+	+	+	+	−	+	+	−	+
16	+	−	−	−	−	−	−	−	−	+
17	+	+	+	+	−	−	+	+	+	+
18	−	−	+	−	+	+	−	−	+	+
19	+	+	+	+	+	+	+	+	+	+
20	+	+	+	+	+	+	+	+	+	−

in Table 2.2.1, we obtain the 10 × 10 proximity matrix given in Table 2.2.2. Each entry of the matrix represents the similarity between the corresponding pair of tests.

Alternatively, if we were interested in the similarities among the 20 chemicals, we could compute the number of matching results between each pair of chemicals and divide by the total number of measurements. This would produce a 20 × 20 proximity matrix where each entry would represent the similarity between the corresponding pair of chemicals. The construction of this matrix will be left to the reader.

Once the raw data has been transformed into either a normalized pattern matrix or a proximity matrix, an appropriate clustering method (or methods) can be used to analyze which groups of data items most resemble one another.

2.2.2. Clustering Methods

If we have a pattern matrix containing ratio-scaled measurements, the class of methods called "partitional clustering methods" can be applied. As noted earlier, partitional methods divide or "partition" the data items into natural groups or clusters based on similarities in the data patterns. A typical algorithm in this class of methods would proceed as follows: First, the analyst would select the number of groups among which the algorithm should divide the data items. (Note: Generally this number is not known in advance, but changes with successive applications of the algorithm until the "best number of partitions" is obtained.) The algorithm would then assign each data item to one of the groups. A measure of the "closeness" or similarity of the data items in each group would then be computed. If the similarity of the data items in any of the groups is not very high, then

TABLE 2.2.2. Proximity Matrix Computed from Table 2.2.1 Using Matching Coefficients

	T_1	T_2	T_3	T_4	T_5	T_6	T_7	T_8	T_9	T_{10}
T_1	1.0	0.65	0.60	0.90	0.60	0.45	0.80	0.80	0.65	0.65
T_2	0.65	1.0	0.85	0.75	0.75	0.60	0.85	0.85	0.70	0.60
T_3	0.60	0.85	1.0	0.70	0.80	0.65	0.80	0.80	0.65	0.55
T_4	0.90	0.75	0.70	1.0	0.60	0.55	0.90	0.90	0.65	0.65
T_5	0.60	0.75	0.80	0.60	1.0	0.65	0.70	0.70	0.65	0.55
T_6	0.45	0.60	0.65	0.55	0.65	1.0	0.45	0.55	0.80	0.60
T_7	0.80	0.85	0.80	0.90	0.70	0.45	1.0	0.90	0.55	0.65
T_8	0.80	0.85	0.80	0.90	0.70	0.55	0.90	1.0	0.65	0.55
T_9	0.65	0.70	0.65	0.65	0.65	0.80	0.55	0.65	1.0	0.55
T_{10}	0.65	0.60	0.55	0.65	0.55	0.60	0.65	0.55	0.55	1.0

Fundamental Basics of the CPBS Approach

the algorithm would select one or more of the data items to change their group membership in order to improve the closeness indices of the groups. The algorithm would stop either after a preset number of improvement iterations or after a level of closeness has been achieved, or when additional iterations result in very little improvement in the level of closeness. Partitional clustering methods differ in (1) how the algorithm assigns group membership, (2) how the algorithm measures closeness, (3) how the algorithm changes group membership, and (4) the stopping criterion.

The other types of clustering methods are applied to proximity matrices. These are called the hierarchical clustering methods. Hierarchical methods produce a nested sequence of groupings. A typical algorithm in this class would proceed as follows: Initially, the algorithm would place each data item in a separate cluster or group. Then in each proceeding iteration, a new cluster would be formed from two old clusters based on the measures of alikeness found in the proximity matrix and the algorithm's criterion for "linking" or combining clusters. The algorithm would continue to link pairs of clusters until all data items are in one large cluster. Hierarchical methods differ mainly in the criterion used to link the clusters.

As we will see, the results of the hierarchical results can be displayed graphically. This manner of display provides an idea of which data items group together and also some idea of how strongly they group together. The following are more detailed discussions and examples of the three hierarchical methods that we have employed in the CPBS.

2.2.3. Single-Link Method

The single-link method begins by setting an initial proximity level at its maximum value and assigning each data item to a separate cluster. At each iteration, a lower proximity level is considered and the current clusters are examined to see whether any of them should be merged based on the "single-link criterion." The criterion used for "linking" two clusters is that two clusters are merged if *at least one pair* of data items (constructed by taking one data item from each cluster) has a proximity measure at or above the current proximity level (hence the name "single link"). At the final iteration (where the proximity level is at its minimum) all pairs of data items have proximities at or above this level, thus all data items are merged into one cluster.

As an illustration of the single-link method, consider the 10×10 proximity matrix given in Table 2.2.2. The results for the single-link method on this proximity matrix are given in Table 2.2.3 and Fig. 2.2.3.

Looking at the proximity matrix given in Table 2.2.2 and Fig. 2.2.3, note that at the 0.9 proximity level, tests T_1, T_4, T_7, and T_8 form a single

TABLE 2.2.3. Results of the Single-Link Method on the Proximity Matrix Given in Table 2.2.2

Proximity level	Clusters
1.00	(T_1) (T_2) (T_3) (T_4) (T_5) (T_6) (T_7) (T_8) (T_9) (T_{10})
0.90	(T_1, T_4, T_7, T_8) (T_2) (T_3) (T_5) (T_6) (T_9) (T_{10})
0.85	$(T_1, T_2, T_4, T_7, T_8)$ (T_3) (T_5) (T_6) (T_9) (T_{10})
0.80	$(T_1, T_2, T_3, T_4, T_5, T_7, T_8)$ (T_6, T_9) (T_{10})
0.75	$(T_1, T_2, T_3, T_4, T_5, T_6, T_7, T_8, T_9)$ (T_{10})
0.65	$(T_1, T_2, T_3, T_4, T_5, T_6, T_7, T_8, T_9, T_{10})$

cluster. Note also that the pairs (T_4, T_7), (T_7, T_8), and (T_4, T_8) are each at the 0.9 proximity level; however, for test 1, the pair (T_1, T_4) is at the 0.9 proximity level, but the pairs (T_1, T_7) and (T_1, T_8) are not (they each have 0.8 proximities). This demonstrates the requirement that in order for a cluster to be merged with another cluster, it need only have one data item with a proximity at the given level (in this case 0.9) to one of the data items in the other group.

As a further illustration of this "linking" criterion, consider T_2 at the 0.85 proximity level. Note that T_2 merges with the cluster (T_1, T_4, T_7, T_8) because it has at least one "link" to the group at the 0.85 proximity level. (Actually it has two: namely, $[T_2, T_7]$ and $[T_2, T_8]$.)

2.2.4. Spanning Tree Method

The spanning tree method is similar to the single-link method in that it produces a "single link" between a given data item and the one that is in closest proximity to it. The two methods differ in the manner in which

```
Proximity
  Level    T1   T4   T7   T8   T2   T3   T5   T6   T9   T10

  1.00     |    |    |    |    |    |    |    |    |    |
  0.95     |    |    |    |    |    |    |    |    |    |
  0.90     |--------------|    |    |    |    |    |    |
  0.85              |--------------|    |    |    |    |
  0.80                   |-------------|    |-----|    |
  0.75                          |--------------|       |
  0.70                                 |               |
  0.65                                 |---------------|
```

FIGURE 2.2.3. Graphical display of the single-link results given in Table 2.2.3.

Fundamental Basics of the CPBS Approach

the results are displayed. The spanning tree method provides a view of the structure of the data that is slightly different from that of the single-link method.

The diagram produced by the spanning tree method shows which data items are linked together, as well as the strengths of the linkages (i.e., the proximities). For example, consider the same proximity matrix used in the previous example (Table 2.2.2). If the spanning tree method is applied to these data, the results given in Fig. 2.2.4 are obtained.

Note that the spanning tree diagram does not provide the hierarchical view of the proximities (i.e., it does not show which clusters formed at a particular proximity level) as does the single-link method, but it does clearly show which tests are in closest proximity to each other, which the single-link diagram does not. Thus the two methods together provide a clearer "picture" of the discordance and concordance among pairs of data items.

2.2.5. Complete-Link Method

The complete-link method uses a different (and more restrictive) criterion for linking two clusters. As in the single-link method, the complete-link method begins with each data item in a separate cluster and initializes the proximity level at its maximum value. Each iteration of the method considers a lower proximity level and merges two clusters if *all pairs of data items* have proximities at or above the current level (hence the name "complete link").

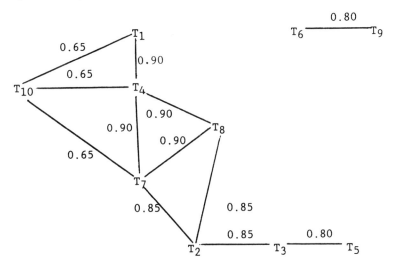

FIGURE 2.2.4. Graphical display of spanning tree results.

TABLE 2.2.4. Results of the Complete-Link Method on the Data Given in Table 2.2.2

Proximity level	Clusters
1.00	(T_1) (T_2) (T_3) (T_4) (T_5) (T_6) (T_7) (T_8) (T_9) (T_{10})
0.90	(T_1, T_4) (T_7, T_8) (T_2) (T_3) (T_5) (T_6) (T_9) (T_{10})
0.85	(T_1, T_4) (T_2, T_3) (T_7, T_8) (T_5) (T_6) (T_9) (T_{10})
0.80	(T_1, T_4, T_7, T_8) (T_2, T_3) (T_6, T_9) (T_5) (T_{10})
0.75	(T_1, T_4, T_7, T_8) (T_2, T_3, T_5) (T_6, T_9) (T_{10})
0.60	$(T_1, T_4, T_7, T_8, T_2, T_3, T_5)$ (T_6, T_9, T_{10})
0.45	$(T_1, T_4, T_7, T_8, T_2, T_3, T_5, T_6, T_9, T_{10})$

Again consider the same proximity matrix. The results of the complete-link method on this data are given in Table 2.2.4 and Fig. 2.2.5. Notice that in contrast to the single-link method, the complete-link method merges the tests T_1, T_4, T_7, and T_8 at the 0.8 proximity level rather than the 0.9 proximity level. This is because the pairs of tests (T_1, T_7) and (T_1, T_8) are only 80% similar and the complete-link method requires that *all pairs* of tests must have proximities at the given proximity level in order for them to be merged at that level.

It should be noted that in the case of "ties" (i.e., when two or more pairs of tests have the same proximities), the order in which the clusters are formed can make a difference in the results that are obtained. For example, you could get clusters (T_1) and (T_4, T_7, T_8) rather than (T_1, T_4) (T_7, T_8) at the 90% level if you did not cluster T_1 to T_4 first. In such cases it is a good idea to do the analysis for each possible ordering in order to

```
Level     T1    T4   T7    T8    T2    T3    T5    T6    T9    T10
1.00      |     |    |     |     |     |     |     |     |     |
0.95      |     |    |     |     |     |     |     |     |     |
0.90      |-----|    |-----|     |     |     |     |     |     |
0.85      |     |    |     |     |-----|     |     |     |     |
0.80      |---------|       |          |     |     |-----|     |
0.75      |          |                 |-------|         |     |
0.70      |          |                         |         |     |
0.65      |          |                         |         |     |
0.60      |------------------|                 |---------------|
0.55                 |                                   |
0.50                 |                                   |
0.45                 |-----------------------------------|
```

FIGURE 2.2.5. Graphical view of the results given in Table 2.2.4 proximity.

Fundamental Basics of the CPBS Approach 41

see whether the order in which the clusters are formed has any major impact on the results. Note that, in this example, the results would be the same (i.e., the same clusters would be obtained) at the 80% level regardless of the ordering. Thus for this specific example, there is no major impact on our results due to the presence of ties in the proximity matrix.

2.2.6. Interpretation of Clusters

In the single-link method we focus on the level of proximity and observe which data items form a cluster at that level. Clusters formed by this method are "weak." In other words, if we were to look at the results of this analysis on the example (see Fig. 2.2.3) and were to focus on the 90% similarity level, the only thing we could conclude is that *some* of the tests within the clusters at that level are 90% similar. For example, for the cluster (T_1, T_4, T_7, T_8) produced at the 90% similarity level, we cannot conclude that all of the tests are 90% similar; the only requirement is for one pair of tests within the given cluster to be 90% similar.

As noted earlier, the spanning tree method is based on the same criteria for forming clusters as the single-link method. Although both methods have the same theoretical and mathematical basis, the results are displayed in different formats and consequently each provides a different perspective of the data. In contrast to the single-link method, the spanning tree method stresses pairwise relationships between objects rather than groupings at various levels of similarity. As was true with the single-link method, the clusters pictured by the spanning tree method are weak and should be interpreted with care. For example, in the spanning tree diagram displayed in Fig. 2.2.4, it appears that the group of tests T_4, T_7, T_8, and T_2 form a strong cluster at the 85% similarity level; it should, however, be noted that the similarity between T_2 and T_4 is only 75%. The only sure relationships are the ones displayed in the diagram (in other words, transitivity of similarities does not hold).

With the complete-link method, we obtain much stronger clustering than with either of the previous methods. As in the single-link method, the complete-link method also highlights the groupings at various levels of proximity; but in contrast to both of the previous methods, if we look at a cluster formed at the 90% proximity level, we are assured that each member of the cluster is at least 90% similar to *all* other members within that cluster. It is this "strength" of the clusters that gives us the most confidence in the clusters displayed by the complete-link method. But this same strength can hide some of the strong pairwise similarities.

In the CPBS approach, we suggest the use of all three methods. Each method gives a slightly different perspective of the data and consequently we should be able to obtain a more complete picture of the relationships in the data base if all three methods are used.

2.2.7. Verification/Validation of Clusters

The formation of clusters by any of the three previously described methods is strongly related to the choice of similarity or proximity measure used. For our previous example, the similarity between two tests was measured as the percentage (or ratio) of matching test results. Thus, it is a measure of the similarity between the *responses* of the tests rather than a direct measurement of the tests themselves. Information about modes of action, mechanisms, etc., is not used in the formation of the clusters.

Verification and validation of clusters involves closely examining the clusters and trying to determine whether the results characterize the data. It consists of following objective procedures to verify whether the data structure formed by the clustering method is strong enough to provide statistical evidence about the phenomenon being studied.

In practice, users of cluster analysis are usually satisfied when intuitive indicators of validity are satisfied (e.g., when the user is merely looking for evidence to support prior hypotheses about the data). For example, when cluster analysis is applied to short-term test data, one might have an *a priori* hypothesis that tests using the same biological end points (e.g., gene mutation assays, Salmonella assays) would form groups. In this case, validation of clusters involves the use of creativity, experience, and insight into the phenomenon under study to try to explain why particular clusters formed. An example of this type of cluster verification is given in Section 3.3.

What happens when cluster analysis is truly being used as an exploratory tool and some of the clusters produced by the methodology lack explanation? In this case the analyst must determine whether these unexplained clusters are real or simply an artifact of the method—no matter what data, fictitious or real, are given to a cluster analysis method, the method will produce clusters! A number of validation procedures are described in the literature. For example, Ling (1972, 1973) calls a cluster real if it forms early in the single-link method and remains separate for a reasonably long range of proximities. Gnanadesikan *et al.* (1977) suggest the use of "jackknifing," where one would omit a measurement from the analysis and investigate the effects on the clustering structure: the structure should remain fixed through minor changes in the data. Smith and Dubes (1979) suggest randomly dividing the data in half and comparing the results of cluster analysis. Here, if the clusters are "real," the results should be approximately the same on both halves of the data. Many other procedures

Fundamental Basics of the CPBS Approach

for testing the validity of clusters are also described in the literature (see Dubes and Jain, 1979, for a more complete discussion of validity).

2.2.8. Closing Remarks

As noted at the beginning, cluster analysis is an exploratory data analysis tool. It provides qualitative verification about natural groupings in the data and can also lead to the initiation of new hypotheses. If valid clusters are found that lack explanation, this can prod one's creativity to delve more deeply into the underlying problem, ultimately leading to new research and new ideas. Specific descriptions of how cluster analysis is used in the CPBS are given in Chapters 3 and 4. For more general information on cluster analysis, the reader is referred to the references, such as the book by Anderberg (1973).

2.3. MULTIPLE-OBJECTIVE DECISION MAKING

The selection and use of multiple tests in industrial and regulatory decision making often involves the consideration of multiple, noncommensurate, and often conflicting objectives. For example, when trying to determine what battery of tests would be "best" to use, one might consider the dollar costs of the various tests to be important, as well as the need to obtain the test results rapidly and the need for accurate results. The problem then becomes one of minimizing cost, minimizing the time required for testing, and maximizing the reliability of the test results. These objectives are each given a different measurement scale, and consequently it may not be clear how to compare the objectives or to give weights to them. In addition, the lowest cost battery would generally not be the one providing the best performance or reliability. Thus we are also left with the problem of trading-off one objective over others.

During the past two decades the explicit consideration of multiple objectives in modeling and decision making has grown by leaps and bounds. The 1970s in particular have seen the emphasis shift from the dominance of single-objective modeling and optimization toward an emphasis on multiple objectives. This has led to the emergence of a new field that has come to be known as multiple-criteria decision making (MCDM). Since the consideration of multiple objectives is an integral part of decision making when multiple tests are available, the aim of this section is to provide a brief overview of MCDM to the reader, and to describe and illustrate some of the basic concepts and procedures that are used. For more details on this topic, the reader is referred to Chankong and Haimes (1983), Yu (1985), and Steuer (1986).

2.3.1. Multiple-Criteria Decision Making

MCDM has emerged as a philosophy that integrates common sense with empirical, quantitative, normative, descriptive, and value-judgment-based analysis. It is a philosophy supported by advanced systems concepts (e.g., data management procedures, modeling methodologies, optimization and simulation techniques, and decision-making approaches) that are grounded on both the arts and the sciences for the ultimate purpose of improving the decision-making process.

The term *multiobjective decision making* refers to the entire process of problem solving, consisting essentially of the five steps depicted in Fig. 2.3.1. The process begins when the decision maker perceives the need to alter the course of the system about which he/she is concerned. For example, in the process of developing a new product, the manager of a chemical company may perceive the need to evaluate the potential cancer hazard of the product in order to determine whether to continue the development of the product and/or whether any special safeguards are needed. Once the situation is diagnosed, general statements of overall needs or objectives are stated and the problem-formulation step begins.

The various tasks in problem formulation normally include (1) translating vaguely stated overall objectives into a more operational set of specific multiple objectives and (2) clearly specifying all essential elements in the system (or problem), system boundary, and system environment. For example, the aforementioned chemical company may have a choice of doing the testing in-house or contracting with an outside laboratory. In addition, there may be a number of different types of short-term tests available. The company's problem may then be to select the "best" testing option (location and composition of testing battery) that would minimize testing costs, maximize the efficiency of the testing process, and maximize the reliability of the results.

Once the system, its environment, and the set of objectives are well defined, appropriate models are constructed. By a "model" we mean a collection of key variables and their logical (or physical) relationships, which together facilitate effective and meaningful comprehensive analysis of the pertinent aspects of the system. In general there are various forms of models, including simple mental models, graphic models (e.g., a chart), complex physical models, and mathematical models. These models have several functions, one of which is to provide the information required to generate available alternative courses of action. In terms of "our company's"

Fundamental Basics of the CPBS Approach

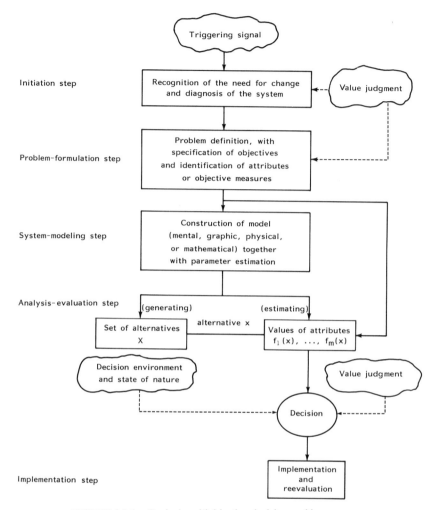

FIGURE 2.3.1. Typical multiobjective decision-making process.

problem, this step might involve processing past performance data on the available tests to evaluate the reliability of the tests, processing data on the costs and time-efficiencies of the two options—in house and outside contract, and analyzing the relationships between the tests (e.g., statistical dependencies and biological relevance).

Once the system has been modeled, we enter the analysis and evaluation step. This step involves generating a set of feasible options and then comparing the set of alternatives. In our problem, the set of alternatives would consist of all feasible testing options. In order to compare the set of alternatives, a set of attributes or objective measures must be clearly specified. The values of these attributes for a given alternative can be obtained either from models or through subjective judgments. The levels of these attributes serve as yardsticks by which the degree of attainment of the objectives specified in the preceding step can be assessed. In our example, this would involve quantifying the costs, time-efficiencies, and reliabilities of each of the feasible testing options.

To complete the analysis and evaluation step, each alternative is evaluated relative to others in terms of a decision rule or set of rules that the decision maker has prespecified in order to rank the available alternatives. The alternative having the highest rank according to the decision rule is then chosen for implementation. This ranking could, for example, be determined through the use of Bayes' rule (see Section 2.1 on Decision Analysis) or utility values.

The process may end here, in which case it is known as an *open-loop process*. Alternatively, if the current results are found to be unsatisfactory, we may use information about the observed output (obtained from implementing the chosen alternative) and return to the problem formulation step. We then speak of a *closed-loop process*. For example, after a testing option has been chosen by the chemical company and the new product has been tested by the chosen battery of short-term tests, the results of the tests may or may not be conclusive enough to make the decision to continue the development of the new product. If the results are conclusive (in the eyes of the decision maker) then we have an open-loop process and the company can make the decision to continue or discontinue the development of the product. Alternatively, if the results are not conclusive then we have a closed-loop process and the company must decide which additional tests should be used. This requires going back to the problem formulation stage, since the goals and objectives of the "additional testing problem" might be different from the original problem's goals.

2.3.2. Basic Elements of Multiple-Objective Problems

In every decision process and in any decision situation, there are *factual elements* and *value elements*. Factual elements are those that can be verified scientifically and subjected to scientific processes that lead to other verifiable elements. For example, in the problem of using multiple tests for decision making, factual elements would include the performances and dependencies among the tests (that can be verified through the use of statistical models) as well as the costs of the tests.

Value elements, on the other hand, defy all forms of scientific verification and treatment. A collection of value elements and their sources

Fundamental Basics of the CPBS Approach 47

constitutes *the value system*. *Judgment*, which signifies the act of giving an opinion, is the most common value element in any decision-making process. The judgment that it is more important to keep the cost of testing down than to keep the time it takes to do the testing down is an example of a value element.

There has been debate on the emphasis and extent to which a value system, particularly the element of judgment, should be incorporated in the actual decision-making process [see, for example, the preface of Easton's book (1973) or papers by Black, 1975]. On the one hand, there are those who tend to ignore value judgments altogether and who justify the "rationale" of this approach by claiming that value systems are the main source of "irrationality." A hard-line manager or administrator, on the other hand, may make some decisions based solely on intuition, experience, and good judgment. It is not our intention here to weigh the arguments. What we will attempt to accomplish here is to identify how and where judgment is required in the multiobjective decision-making process. Indirectly, therefore, we will also be able to identify those tasks that can be carried out through formal procedures. In this way it is hoped that we can strike a proper balance between objectivity and subjectivity in solving multiobjective decision problems involving the use of multiple tests.

An example of a task in the multiobjective decision-making process (Fig. 2.3.1) in which subjective judgment is an integral part would be in the initiation step, where recognizing the need for change and perceiving the overall objectives to be achieved are purely subjective processes, involving soul-searching for "needs" and "wants." Other tasks in the multiobjective decision-making process that are mostly subjective include (1) defining clearly the statement of the problem, the system boundary, and the system environment in the problem formulation step; (2) articulating the set of specific objectives and identifying appropriate attributes in the problem formulation step; (3) choosing the type of model to be used in the modeling step; (4) choosing key variables to be incorporated in the model in the modeling step; and (5) choosing the set of decision rules in the analysis and evaluation step.

An example of a task that may require value judgments is the development of logical relationships among the chosen decision variables, as well as between variables and the chosen attributes in the modeling step. This development may be either purely subjective, if only mental models are involved, purely analytical or quantitative, in cases where mathematical models are appropriate representations of the system, or a mixture of both if some other forms of models are used. Other tasks requiring value judgments would be (1) evaluating the impact of each alternative, i.e., calculating the value of each attribute and assessing the relationship between attributes and objectives in the analysis and evaluation step, and (2) applying the decision rule to arrive at the final decision in the analysis and evaluation step.

Attempts have been made to deal with the judgments involved in these tasks. Experimental psychologists and behavioral scientists seek to understand through empirical studies the psychology underlying judgment. Management scientists, operations researchers, and systems engineers, on the other hand, are more interested in finding appropriate guidelines for effectively combining judgment with some formal procedures. Perhaps the most successful treatment of judgment in multiobjective decision problems is in the area of the preference structure of the decision maker. This type of judgment is required in the evaluation step, outlined in Fig. 2.3.1. Unidimensional utility theory, multiattribute utility theory, and other related topics all represent attempts to formalize the preference structure of the decision maker.

Another element of any multiobjective decision problem is the *decision-making unit and the decision maker.* As noted by Rigby (1964), the term "decision maker" is difficult to define precisely. Churchman (1968) refers to the decision maker as "the person who has the ability to change the system, i.e., the responsibility and authority for such a change." More specifically, and more appropriate to the context with which we shall be concerned, we shall take the *decision maker* to be an individual or a group of individuals who directly or indirectly furnish the final value judgment that may be used to rank available testing alternatives (so that the "best" testing choice can be identified); and the individual or group of individuals (which may or may not be the same as the preceding) who furnish the value judgment on whether particular testing results are conclusive enough to base their operating decisions on. Implicit in this statement is that whenever final value judgments need to be made concerning the "goodness" or "badness" of a given choice, they are to be made by the decision maker.

A *decision-making unit* consists of the decision maker and the support staff and "machinery," which together act as an information processor to perform the following functions: (1) receive input information, (2) generate information within itself, (3) process information into intelligence, and (4) produce the decision. Thus the smallest decision-making unit would be the decision maker. A larger unit may, for example, consist of the decision maker, system analysts, statisticians, and computing and graphic instruments.

2.3.3. Basic Procedures Used in Multiobjective Decision Making

For notational convenience, let us define the general multiple-objective optimization problem to be

Fundamental Basics of the CPBS Approach

$$\min\{f_1(\mathbf{x}), f_2(\mathbf{x}), \ldots, f_m(\mathbf{x})\} \quad (2.3.1)$$

subject to

$$g_i(\mathbf{x}) \leq 0, \quad i = 1, 2, \ldots, k$$

where $\mathbf{x}$ is an n-dimensional vector of decision variables; $f_i(\mathbf{x})$, $i = 1, 2, \ldots, m$, are m objective functions; and $g_i(\mathbf{x})$, $i = 1, 2, \ldots, k$, are k constraint functions. For example, the decision variables ($\mathbf{x}$) could represent the tests that are under consideration; the objectives [$f_i(\mathbf{x})$] could be to minimize cost, to minimize false negative results, and to minimize false positive results; and the constraints could deal with the maximum allowable time for testing and the biological requirements for battery composition.

It is important to emphasize that a mathematical optimum to the model of a decision problem such as that given by Eq. (2.3.1) may exist, but the optimum to a real-life decision problem *does not exist in an objective sense per se*. An "optimum" solution to a real-life problem depends on myriad factors, which include such factors as who the decision makers are, what their perspectives are, what the biases of the modeler are, what the credibility of the data base is. Therefore, a mathematical optimum to a model such as that given in Eq. (2.3.1) does not necessarily correspond to the optimum for the real-life problem.

In addition, for problems where more than one objective is important, a single solution to the model of the decision problem given in Eq. (2.3.1) with $m \geq 2$ may not exist. For example, in the problem of finding a battery that has both minimum cost and maximum performance, one may not be able to find such a battery of tests. Instead we may find that some combinations of tests have a low cost but relatively poor performance, whereas other combinations may have very good performance but high costs. This necessitates the concept of *noninferior solutions*.

The concept of noninferior solutions, also known as Pareto optimum, efficient solution, etc., finds its basic roots in economics in general and in competitive equilibrium in particular (Koopmans, 1951; Intriligator, 1971). Koopmans defined an efficient point for multiobjective functions in economics as follows: "A possible point in the commodity space is called efficient whenever an increase in one of its coordinates (the net output of one good) can be achieved only at the cost of a decrease in some other coordinate (the net output of another good)." Kuhn and Tucker (1950) extended the theory of nonlinear programming for one objective function to a vector optimization problem and introduced necessary and sufficient conditions for a "proper" solution to Eq. (2.3.1). It has been shown (under some mild restrictions on decision-maker preference structure) that the decision maker's "optimal" solution will be one of the noninferior solutions (see, for example, Chankong and Haimes, 1983; Yu, 1985).

A formal definition of a noninferior solution can be given as follows: A decision x^* is said to be a noninferior solution to the problem posed by the model given in Eq. (2.3.1) if and only if there does not exist another feasible x such that $f_i(x) \leq f_i(x^*)$, $i = 1, 2, \ldots, m$, strict inequality holding for at least one i. Consequently, any alternative at which no one of the objective functions $f_i(x)$ can be improved without causing a degradation in any other $f_j(x)$, $i = j$, is a noninferior solution of Eq. (2.3.1). In general, more than one noninferior solution will exist; and not all feasible alternatives [i.e., those solutions satisfying the constraints given in Eq. (2.3.1)] will be noninferior. Thus one approach to finding the solution to Eq. (2.3.1) would be to first find all of the noninferior solutions and then to incorporate the value judgment of the decision maker to select the "best" option. (Here the term "best" is defined in terms of *trade-offs* among the objectives.) One such methodology—the surrogate worth trade-off (SWT) method and its extensions (Haimes and Hall, 1974; Haimes *et al.*, 1975; Chankong and Haimes, 1983) can be used for this purpose.

In the following discussion, we illustrate some of the preceding concepts through the use of two examples involving the use of multiple tests for decision making. In our first example, suppose we are interested in obtaining more accurate information about the ability of short-term tests to identify cancer hazards. The ability of a test to identify the carcinogenic potential of a chemical can be characterized by the test's sensitivity and specificity. These measures are determined from past data on the test. Since the accuracy of these measures increases as more chemicals with known carcinogenicity are tested, and, as a corollary, the associated cost of testing will increase with it, there are definite trade-offs between (1) obtaining accurate estimates of the sensitivity and specificity measures and (2) the cost of additional testing.

It is mathematically as well as graphically more convenient to analyze and display trade-offs between two objectives when both objectives are either minimized or maximized simultaneously. We will, therefore, keep the minimization of cost as is, and convert the maximization of the accuracy of sensitivity or the accuracy of specificity into a minimization problem. Clearly, increasing the accuracy of these performance measures by increasing the number of objects tested will decrease their associated variances.

Let $f_1(x)$ represent the cost of testing, $f_2(x)$ represent the variance of the computed sensitivity, and $f_3(x)$ represent the variance of the computed

Fundamental Basics of the CPBS Approach

specificity, where **x** is the number of chemicals tested. Thus the optimization problems for a particular test can be stated as:

$$\min\{f_1(\mathbf{x}), f_2(\mathbf{x}), f_3(\mathbf{x})\} \quad (2.3.2)$$

where (2.3.2) is subject to $\mathbf{x} \in X$, where X denotes the set of all feasible values for the number of chemicals tested (e.g., the company may have available to it 20 known carcinogens and 20 known noncarcinogens).

Figure 2.3.2 depicts a generic representation of the noninferior solutions for two of the objectives $[f_1(\mathbf{x})$ and $f_2(\mathbf{x})]$ in the problem posed by Eq. (2.3.2). Points A, B, C, and D represent four different noninferior testing policies. The objectives in Fig. 2.3.2 call for minimizing both the cost of testing and the variance of the sensitivity index. Note that testing policy A is associated with a very high cost (a large amount of additional testing), but the sensitivity would have a low variance (high accuracy); on the other hand, testing policy D is associated with a low cost (low amount of additional testing) but it has a high variance (low accuracy). The trade-off at policy A is very high; namely, with a small marginal decrease in the accuracy (increasing the variance), a major reduction in the testing cost can be achieved. Policy D represents yet another extreme. With a minor marginal increase in the testing cost, a substantial increase in the accuracy of the sensitivity index can be achieved. Reasonable decision makers may opt to move from extreme policy A toward B and C and from extreme policy D

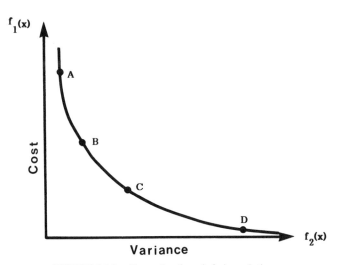

FIGURE 2.3.2. Example of noninferior solutions.

to C and B. The best-compromise solution (policy) is more likely to be reached in the neighborhood of B or C than in the neighborhood of A or D.

As a second illustration, consider the situation in which a chemical company has just completed a battery of tests on a chemical with unknown carcinogenic potential. Suppose that the current results were inconclusive, and that the choice must be made as to whether additional testing should be used, and if so what tests should be used. We could formulate the problem (model the problem) in terms of minimizing the cost $f_1(\mathbf{x})$ of the additional test(s) and minimizing the rate of getting false positives $f_2(\mathbf{x})$ (which can be given by one minus the specificity) subject to $x \in X$, where X denotes the set of all feasible batteries of tests. Table 2.3.1 and Fig. 2.3.3 depict a representation of seven feasible battery options and illustrate which of them would be noninferior. [The example is based on data from actual short-term tests given in Rosenkranz *et al.* (1984) and Pet-Edwards *et al.* (1985).]

The points A, B, C, G, and D represent five different noninferior testing policies. The objectives in Fig. 2.3.3 call for minimizing both the cost of testing and the false positive rate. Note that testing policy A is associated with a very high cost of testing but with a low false-positive rate, whereas testing policy D is associated with very low cost and a high false-positive rate.

Clearly, policies (test batteries) E and F are inferior solutions, since one can always find better policies. For example, testing policies E and C yield the same false-positive rate of 0.3; however, policy E costs $600 and policy C costs only $200. A similar argument can be made for policies B and F, which cost the same ($400); however, policy B yields a low false-positive rate of 0.14 and policy F yields a much higher false-positive rate of 0.53.

TABLE 2.3.1. An Illustration of Feasible Solutions, Noninferior Solutions, and Trade-offs Among Noninferior Solutions

Battery	$f_1(x)$[a]	$f_2(x)$[b]	Trade-off[c]	Noninferior?
A	$700	0.06	3750	Yes
B	400	0.14	1250	Yes
C	200	0.30	312.5	Yes
D	40	0.74	154	Yes
E	600	0.30		No
F	400	0.53		No
G	100	0.62	500	Yes

[a] $f_1(x)$, the cost of the battery.
[b] $f_2(x)$, the rate of false positives.
[c] Trade-offs are given as dollars versus change in $f_2(x)$.

Fundamental Basics of the CPBS Approach

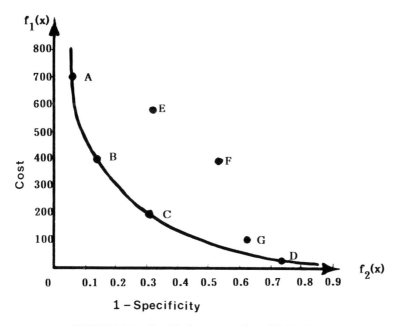

FIGURE 2.3.3. Graphical representation of Table 2.3.1.

2.3.4. Closing Remarks

In most problems involving the use of multiple tests for decision making, there will be many more options available than the seven in our second illustration. From both a theoretical and computational viewpoint, it is possible to consider and evaluate thousands of policies using the approach of multiobjective decision making. The use of trade-offs, such as in column 4 in Table 2.3.1, and figures, such as Figs. 2.3.2 and 2.3.3, are very useful in assisting the decision maker to determine his best-compromise solution(s). Various other approaches to finding the best-compromise solution from among a set of noninferior solutions exist in the literature. For further information, the reader is referred to the following texts: Yu (1985), Chankong and Haimes (1983), and Steuer (1986).

2.4. DYNAMIC PROGRAMMING

The use of multiple tests in decision making often involves the consideration of more than one objective. For example, in selecting an optimal testing strategy, one may be interested in both the costs of the various strategies and how well the strategies are expected to perform. As described in the preceding section, such problems can be "solved" by first finding all of the noninferior solutions and then having the decision maker look at graphs of the results and trade-offs among the objectives in order to facilitate the selection of his most-preferred solution. A considerable effort can be expended in the process of finding all of the noninferior solutions. We have found that a method based on *dynamic programming* can sometimes prove to be a useful tool for generating the noninferior batteries of tests.

Since dynamic programming may be unfamiliar to the reader, in this section we will provide a general discussion on some of the basic concepts underlying dynamic programming. It should be noted that it is difficult to describe dynamic programming in general terms since the formulation of a "dynamic program" is very dependent on the specific problem it is being applied to. Thus we will present two very simple examples, not related to the test selection problem, in order to illustrate the principles upon which dynamic programming is based and to illustrate the formulation of a dynamic program for a specific problem. The formulation for the battery selection problem will be presented in Chapter 4 of this book.

2.4.1. Introduction

Dynamic programming (DP) is a methodology that under certain conditions is best suited for modeling and optimizing sequential decision-making processes. In a nutshell, DP is based on Bellman's principle of optimality, which states that "An optimal policy has the property that whatever the initial state and the initial decisions are, the remaining decisions must constitute an optimal policy with regard to the state resulting from the first decisions" (Bellman and Dreyfus, 1962). The principle of optimality facilitates the derivation of a recurrence relationship between any two successive states in the form of a recursive equation. Among the conditions that must be satisfied for such a recursive equation to be valid is that the process must be Markovian. The basic property of a Markovian process is that for an n-stage decision-making process, after any number of decisions, say k, the effect of the remaining $(n - k)$ stages of the decision process upon the total objective function will depend only upon the state of the system at the end of the kth decision and the subsequent decisions (Bellman and Dreyfus, 1962).

To appreciate the derivation of the recursive equation in a dynamic programming formulation, it is essential that one understands the meaning and role that state variables, decision variables, and stages play in such a formulation. A sequential decision-making process can sometimes be decomposed into a finite and discrete number of stages. At each stage, the state of the system should provide all the relevant information about the

Fundamental Basics of the CPBS Approach 55

system. At each stage and for different levels of the state variable(s), decisions are made to optimize the set of objective functions. Consider, for example, the optimal battery selection process in terms of dynamic programming. Under the assumption of independence among the tests, the battery selection process can be decomposed into n stages, where at stage 1, only one test can be included in the battery (the trivial stage); at stage 2, up to two tests can be included in the battery; and at stage n, up to n tests can be included in the battery. The state of the system can be represented by the performance of the battery, e.g., the sensitivity or the specificity, and the objective might be to minimize the cost of the battery over all feasible values of the state. Obviously, the state variables are dependent on the decision variables—the tests that have been selected to constitute the battery at each stage.

The details on the use of dynamic programming for battery selection (which is a multiobjective problem) are given in Chapter 4 of this book. The remainder of this section is used to illustrate the basic principles of dynamic programming on single-objective problems and the process of deriving a recursive equation. We utilize a simple network example to illustrate the former, and present a resource allocation example to illustrate the derivation of a recursive equation which relates the state of the system and the decisions from one stage to the next.

2.4.2. A Network Example to Illustrate the Basics of DP

In order to better understand the fundamental basics of dynamic programming, consider the game given in Fig. 2.4.1. The rules of the game

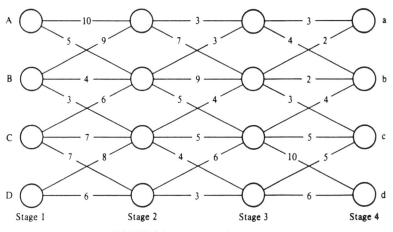

FIGURE 2.4.1. A network example.

are as follows: Start at any point A, B, C, or D and terminate at any destination point a, b, c, or d. The objective is to find the path that will give you the maximum total reward. The reward for each feasible path connecting any two points in the network is shown in the figure. The total reward would be given by the sum of the rewards of the connecting paths. Four stages are designated in the network and it is only possible to proceed from stage i to $i+1$ where a path is shown in the figure.

Clearly, listing all possible paths as a method of solution to this type of combinatorial problem becomes prohibitive for large networks. However, it will be shown that solving this problem via dynamic programming will circumvent the combinatorial calculations and hence effectively reduce the computations needed. This example problem will be used as a vehicle to demonstrate some of the principles upon which dynamic programming is based, and in particular, to demonstrate Bellman's principle of optimality.

The complexity of the problem can be reduced by decomposing the problem into smaller subproblems that are sequentially coupled to one another. In this example, four stages are identified (see Fig. 2.4.1) and a "limited objective function" can be associated with each stage. A stage can be viewed as a subproblem or a subsystem.

At the first stage, we try to answer the following question: Having reached stage two from stage one, what is the maximum reward that can be achieved at each node (circle in the network) and what is the corresponding path? The answer to this question is given in Fig. 2.4.2, where the maximum rewards for each node in the network (written inside the circles of Fig. 2.4.2) are 10, 6, 8, and 7. Often more than one path may yield the

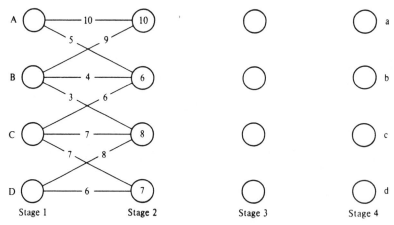

FIGURE 2.4.2. Results of DP on network example from stage 1 to stage 2.

Fundamental Basics of the CPBS Approach

same maximum reward. In that case, more than one solution can be derived. It is important to note that these four maximum reward values replace the ten path rewards that are possible when going from stage one to stage two. In other words, when proceeding from stage two to stage three, there is no need to be concerned with the ten reward values given between stages one and two. Only the four new values in the circle are required to optimally proceed to stage three. This is due to Bellman's principle of optimality.

At the second stage, an answer to the following question is desired: Having reached stage three from stage two, what is the maximum reward that can be achieved at each node (circle in the network) and what is the corresponding path, assuming that an optimal path was chosen in progressing from stage one to stage two. The answer to this question is given in Fig. 2.4.3, where the maximum cumulative rewards for each node in the network (written inside the circles of Fig. 2.4.3) are 13, 17, 13, and 12. For example, advancing from node (A, 2) which had a previous maximum reward of 10 to node (A, 3) yields a cumulative sum of 13. Alternatively, advancing from node (B, 2), which had a previous maximum reward of 6 to node (A, 3) yields a cumulative sum of 9. Clearly, the maximum cumulative reward that can be achieved for node (A, 3) is 13.

Similarly, the maximum cumulative rewards that can be achieved for each of the nodes (A, 4), (B, 4), (C, 4), and (D, 4) are 19, 19, 20, and 23, respectively, as given in Fig. 2.4.4. Obviously, the overall maximum cumulative reward is 23 achieved at node (A, 4) (also designated as d).

In order to find the optimal path, which has resulted in the maximum reward, a backward tracing is necessary. Node (A, 4) was reached from node (C, 3), which in turn was reached from node (C, 2), and finally node

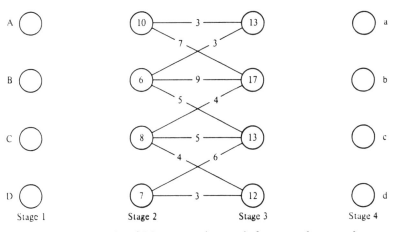

FIGURE 2.4.3. Results of DP on network example from stage 2 to stage 3.

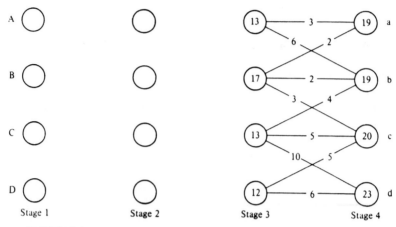

FIGURE 2.4.4. Results of DP on network example from stage 3 to stage 4.

(C, 2) was reached from node (D, 1). The optimal path is therefore, (D, 1) to (C, 2) to (C, 3) to (D, 4) which yields a maximum reward of 23 (see bold path in Fig. 2.4.5).

This example was a simple illustration of the basics of dynamic programming. In particular, our decision variables were the 30 possible partial

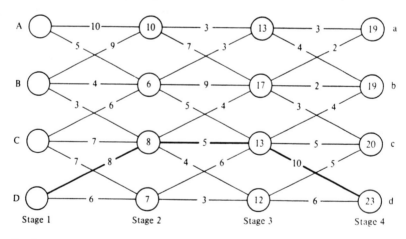

Results:

Maximum reward = 23
An optimal path : (D, 1) to (C, 2) to (C, 3) to (D, 4)

FIGURE 2.4.5. Optimal path for network example.

Fundamental Basics of the CPBS Approach

paths when going from stage i to $(i + 1)$; the three stages were the partial paths of length one, two, and three, respectively; and the state of a partial path was given by the sum of the rewards on that path. The problem was decomposed into three subproblems corresponding to the three stages. In particular, for a given node at stage $(i + 1)$, the stage i subproblem was to maximize the reward for the paths of length i ending at the given node. Because this game satisfied Bellman's principle of optimality, only the results of stage i were required to determine the optimal solution(s) at stage $(i + 1)$. This resulted in a considerable savings in computation. In particular, the dynamic programmming formulation used 12 maximization problems involving 30 paths, whereas the direct computational method would require finding the maximum reward from among 1000 paths.

2.4.3. An Example to Illustrate the Principle of Optimality and Recursive Equations

The network example discussed in the previous section illustrates the basics of dynamic programming. Fundamental to this method was Bellman's principle of optimality. This principle was used in the derivation of a recursive equation that related the state of the system and the decisions from one stage to another.

In the preceding example, we did not attempt to formulate the recursive equations mathematically. As noted earlier, it is difficult to present the dynamic programming recursive equations using general formulas. Thus, the following resource allocation problem will be utilized as a vehicle for demonstrating the procedure used in constructing the recursive dynamic programming equation for a specific problem.

Suppose we are given a resource, in particular water in a reservoir, of capacity Q that can be supplied to N activities, in particular N cities. Let X_i denote the amount of water supplied to the ith city with a return of $g_i(X_i)$. The problem is how many units of water, X_i, to allocate to the ith city in order to maximize the total net return subject to certain constraints. Mathematically, the problem can be formulated as follows: Given

$$N \text{ decision variables:} \quad X_1, X_2, \ldots, X_N$$

and

$$N \text{ return functions:} \quad g_1(X_1), g_2(X_2), \ldots, g_N(X_N)$$

Let $f_N(Q)$ represent the maximum return from the allocation of the capacity Q to the N cities:

$$f_N(Q) = \max_{X_i} \{g_1(X_1) + g_2(X_2) + \cdots + g_N(X_N)\}$$

$$X_i \geq 0, \quad i = 1, 2, \ldots, N$$

subject to a limited quantity Q:

$$\sum_{i=1}^{N} X_i \leq Q$$

It is assumed that the functions $g_i(X_i)$, $i = 1, 2, \ldots, N$, possess the following properties:

1. They are bounded on $0 \leq X_i \leq Q$.
2. They need not be continuous (often these return or cost functions are given in a tabulated or a graphical form).
3. Each $g_i(X_i)$ is a function of only one decision variable.

Note that in this problem there is one state variable only, namely, the water to be allocated to the various cities. The state variable will be represented by q, indicating the amount of resource available for allocation. The number of decision variables is N (the amount allocated to each city).

We can reformulate this problem into stages, where the stage is given by the number of cities (users of water). Note that in this problem the number of decisions is the same as the number of stages. The reader should not conclude that the number of decisions and stages is always the same since this is not always the case.

At the first stage we assume that there is only one potential user of the resource (one city), which will be designated by the subscript 1. Then, since we would still wish to make maximum use of the resource, we define:

$$f_1(q) = \max\{g_1(X_1)\} \qquad (2.4.1)$$

$$0 \leq X_1 \leq q$$

$$0 \leq q \leq Q$$

At the second stage we assume that there are two potential users of the resource (two cities). The new user is designated by the subscript 2. If we allocate to this user an amount X_2, $0 \leq X_2 \leq q$, there will be a return of $g_2(X_2)$ and a remaining quantity of the water $(q - X_2)$ that can be allocated to the first city. Applying the principle of optimality, the optimal

Fundamental Basics of the CPBS Approach

return of the resource for two potential users is
$$f_2(q) = \max\{g_2(X_2) + f_1(q - X_2)\} \tag{2.4.2}$$
$$0 \leq X_2 \leq q$$
$$0 \leq q \leq Q$$

The recursive calculation is now established and $f_3, f_4, \ldots, f_N$ can be written and solved in succession for all possible values of q.

When this process is complete, $f_N(q)$ represents the return from allocating the resource optimally to N users (cities) as a function of the quantity of the resources (amount of water in the reservoir) whatever it may be.

The general recursive relationship of the N activities (stages) is
$$f_N(q) = \max\{g_N(X_N) + f_{N-1}(q - X_N)\} \tag{2.4.3}$$
$$0 \leq X_N \leq q$$
$$0 \leq q \leq Q$$

A direct derivation of the preceding dynamic programming recursive equation is given below following the procedure of Bellman and Dreyfus (1962):

$$\max_{\substack{X_1+X_2+\cdots+X_N=q \\ X_i \geq 0}} = \max_{0 \leq X_N \leq q} \max_{\substack{X_1+X_2+\cdots+X_{N-1}=q-X_N \\ X_i \geq 0}}$$

We can write

$$f_N(q) = \max_{\substack{X_1+X_2+\cdots+X_N=q \\ X_i \geq 0}} [g_N(X_N) + g_{N-1}(X_{N-1}) + \cdots + g_1(X_1)]$$

$$= \max_{0 \leq X_N \leq q} \max_{\substack{X_1+X_2+\cdots+X_{N-1}=q-X_N \\ X_i \geq 0}} [g_N(X_N) + g_{N-1}(X_{N-1}) + \cdots + g_1(X_1)]$$

$$= \max_{0 \leq X_N \leq q} g_N(X_N) + \max_{\substack{X_1+X_2+\cdots+X_{N-1}=q-X_N \\ X_i \geq 0}} [g_{N-1}(X_{N-1}) + \cdots + g_1(X_1)]$$

$$= \max_{0 \leq X_N \leq q} [g_N(X_N) + f_{N-1}(q - X_N)]$$

2.4.4. Closing Remarks

The preceding examples have been given to highlight the fundamentals of dynamic programming and to illustrate the use of Bellman's principle of optimality in the formulation of recursive formulas. The battery selection problem is a multiobjective problem and could benefit from the use of multiobjective dynamic programming (for a discussion of multiobjective dynamic programming see, for example, Yu (1985)]. The formulation for battery selection problem will be given in Chapter 4. For more information on (single-objective) dynamic programming, the reader is referred to Nemhauser (1964).

REFERENCES

Anderberg, M. R., 1973, *Cluster Analysis for Applications*, Academic Press, New York.
Bellman, R. E., and Dreyfus, S. E., 1962, *Applied Dynamic Programming*, Princeton University Press, Princeton, New Jersey.
Berger, J. O., 1985, *Statistical Decision Theory and Bayesian Analysis*, 2nd edition, Springer-Verlag, New York.
Black, M., ed., 1975, *Problems of Choice and Decision*, proceedings of a colloquium in Aspen, Colorado, 24 June-6 July 1974, Cornell University Program on Science, Technology, and Society, Ithaca, New York.
Buchanan, J. T., 1982, *Discrete and Dynamic Decision Analysis*, Wiley-Interscience, Chichester, England.
Chankong, V., and Haimes, Y. Y., 1983, *Multiobjective Decision Making: Theory and Methodology*, Elsevier, North-Holland, New York.
Chankong, V., Haimes, Y. Y., Rosenkranz, H. R., and Pet-Edwards, J., 1985, "The carcinogenicity prediction and battery selection (CPBS) method: A Bayesian approach," *Mutation Res.* **153**:135-166.
Clifford, H. T., and Stephenson, W., 1975, *An Introduction to Numerical Classification*, Academic Press, New York.
Churchman, C. W., 1968, *The Systems Approach*, Dell, New York.
Dubes, R., and Jain, A. K., 1979, *Clustering Methodologies in Exploratory Data Analysis*, Department of Computer Science, Michigan State University, East Lansing, Michigan.
Easton, A., 1973, *Complex Managerial Decisions Involving Multiple Objectives*, Wiley, New York.
Finney, D., 1971, *Probit Analysis*, Cambridge University Press, Cambridge, England.
Gnanadesikan, R., Kettenring, J. R., and Landwehr, J. M., 1977, "Interpreting and assessing the results of cluster analysis," *Bull. Int. Statist. Inst.*, **47**:451-463.
Goldstein, N., and Dillon, W. R., 1978, *Discrete Discriminant Analysis*, Wiley, New York.
Good, I. J., 1978, "Alleged objectivity: A threat to the human spirit," *International Statistical Review*, **46**:65-66.
Haimes, Y. Y., and Hall, W. A., 1974, "Multiobjectives in water resources systems analysis: The surrogate worth trade-off method," *Water Resources Res.*, **10**:615-623.
Haimes, Y. Y., Hall, W. A., and Freedman, H. T., 1975, *Multiobjective Optimization in Water Resources Systems. The Surrogate Worth Trade-Off Method*, Elsevier, New York.
Hamaker, H. C., 1977, "Bayesianism: A Threat to the statistical profession?" *Int. Stat. Rev.*, **45**:111-115.
Holladay, C., 1979, *Decision Making Under Uncertainty: Choices and Models*, Prentice-Hall, Englewood Cliffs, New Jersey.
Intriligator, M. D., 1971, *Mathematical Optimization and Economic Theory*, Prentice-Hall, Englewood Cliffs, New Jersey.
Keeney, R., and Raifia, H., 1976, *Decisions with Multiple Objectives*, Wiley, New York.

Koopmans, T. C., 1951, "Analysis of production as an efficient combination of activities," in *Activity Analysis of Production and Allocation* (T. C. Koopmans, ed.), Wiley, New York, pp. 33-97.

Kuhn, H. W., and Tucker, A. W., 1950, *Contributions to the Theory of Games*, Vol. 1, Princeton University Press, Princeton, New Jersey.

Lindley, D. V., 1985, *Making Decisions*, 2nd edition, Wiley, London.

Ling, R. F., 1972, "On the theory and construction of k-clusters," *Comput. J.*, **15**:326-332.

Ling, R. F., 1973, "Probability theory of cluster analysis," *J. Am. Stat. Assoc.*, **68**:159-164.

McKelvey, R. D., and Zavoina, W., 1975, "A statistical model for the analysis of ordinal level dependent variables," *J. Math. Sociol.*, **4**:103-120.

Meisel, W. S., 1972, *Computer-Oriented Approaches to Pattern Recognition*, Academic Press, New York.

Moore, P. G., 1978, "The mythical threat of Bayesianism," *Int. Stat. Rev.*, **46**:67-73.

Nemhauser, G. L., 1966, *Introduction to Dynamic Programming*, Wiley, New York.

Pet-Edwards, J., Rosenkranz, H. R., Chankong, V., and Haimes, Y. Y., 1985, "Cluster analysis in predicting the carcinogenicity of chemicals using short-term assays," *Mutation Res.*, **153**:167-185.

Raiffa, H., 1968, *Decision Analysis: Introductory Lectures on Choices Under Uncertainty*, Addison-Wesley, Reading, Massachusetts.

Rigby, F. D., 1964, Heuristic analysis of decision situations, in *Human Judgments and Optimality* (M. W. Shelly II and G. L. Bryan, eds.), Wiley, New York.

Rosenkranz, H. R., Klopman, G., Chankong, V., Pet-Edwards, J., and Haimes, Y. Y., 1984, "Prediction of environmental carcinogens: A strategy for the mid 1980's," *Environ. Mutagen.*, **6**:231-258.

Savage, L. J., 1962, "Subjective probability and statistical practice," in *The Foundation of Statistical Inference* (L. J. Savage, ed.), Methuen, London, pp. 9-35 (discussion pp. 62-103).

Savage, L. J., 1954, *The Foundation of Statistics*, Wiley, New York.

Smith, S. P., and Dubes, R. C., 1979, "The stability of hierarchical clustering," Technical Report No. TR-79-02, Computer Science Department, Michigan State University, East Lansing, Michigan.

Steuer, R. E., 1986, *Multiple Criteria Optimization: Theory, Computation, and Application*, Wiley, New York.

Von Neumann, J., and Morgenstern, O., 1953, *Theory of Games and Economic Behavior*, 3rd. edition, Princeton University Press, Princeton.

Weinstein, M. C., Feinberg, H. V., Elstein, A. S., Frazier, H. S., Neuhauser, D., Neutra, R. R., and McNeil, B. J., 1980, *Clinical Decision Analysis*, W. B. Saunders, Philadelphia.

Winkler, R. L., 1972, *Introduction to Bayesian Inference and Decision*, Holt, Reinhart, and Winston, New York.

Yu, P. L., 1985, *Multiple-Criteria Decision Making, Concepts, Techniques, and Extensions*, Plenum Press, New York.

Part II

Carcinogenicity Prediction and Battery Selection Methodology

The carcinogenicity prediction and battery selection (CPBS) methodology is a collection of methods designed to aid in the selection and interpretation of tests used for decision making. It begins with the analysis of a data base containing test results, guides the decision maker in selecting the most preferred battery of tests, and assists the decision maker in interpreting the results of a battery of tests. Thus the CPBS methodology can be described as consisting of three major types of activities: (1) preliminary data analysis (which is used to evaluate and summarize the "information content" of the data base for use in battery selection and Bayesian prediction), (2) battery selection (which is used to help decision makers design and select their most preferred battery of tests), and (3) Bayesian prediction (which helps the decision maker interpret the results of a given battery of tests on an object with unknown properties). It should be noted, however, that not all three activities need to be performed in a specific application. For example, if the performance characteristics of the individual tests and their interrelationships are already well understood, then preliminary data analysis need not be carried out. If the decision maker already knows what battery of tests he/she wishes to use (for example, in the case where there is only one battery available), then the battery selection step may be omitted. The choice of which activities (i.e., preliminary analysis, battery selection, and/or Bayesian prediction) to perform is dependent on the user's interest and the information that is available on the tests.

Part II of this book consists of detailed descriptions of each of the major activities of the CPBS. It is divided into three chapters. Chapter 3 is devoted to a discussion on preliminary analysis, Chapter 4 presents approaches to battery selection, and Chapter 5 discusses how test results can be interpreted using Bayes' theorem.

Chapter 3

Preliminary Analysis

Suppose we would like to utilize the results of multiple tests in making a decision. For example, we might want to use clinical test results in order to diagnose a patient's disease, or we might want to use short-term *in vitro* tests to determine whether a particular chemical would present a cancer hazard. Before we can interpret the results of multiple tests and before we can determine which tests might be appropriate to use, we must have some past data on how well the tests performed in this function. We might already have specific knowledge about the reliability of the individual tests in predicting the property of interest, as well as the interrelationships among the tests. In this case, the analysis of data on the tests (i.e., the preliminary analysis step) may be omitted in its entirety or portions may be skipped.

On the other hand, such direct knowledge about the tests may not be readily available. This could occur, for example, when the group of tests that are being considered have not been specifically used for the type of decision problem under investigation. The use of short-term tests in cancer hazard identification is an example of this situation. In this case, we would have to collect or retrieve the raw test scores on objects having known properties. This collection of test scores will hereafter be termed the *data base*. Then we would need to analyze the reliabilities of the tests and determine their interrelationships; i.e., we would need to perform preliminary analysis of the data in the data base.

The purpose of preliminary analysis is to evaluate and summarize the test data contained in the data base. The results of preliminary analysis will provide reliability and performance indices about the individual tests in a form that is useful for battery selection and prediction. It will also indicate whether more data are needed.

Before we begin our discussion on preliminary analysis, we present the following example in order to highlight and illustrate some aspects of preliminary analysis that are important to the CPBS approach.

An Illustrative Example

Physicians routinely request laboratory tests and obtain physical examination results in order to diagnose their patients' illnesses. These tests and results include, for example, a white blood cell count, a urine culture, temperature, blood pressure, and respiration. Suppose that the presence or absence of a particular disease, D, can be perfectly identified by the result of a blood culture, but that a physician is interested in seeing whether a set of routine clinical tests might prove useful in determining whether his patient has the disease, D, or does not have the disease, ND, since the routine test results can be obtained much more quickly than the blood culture results. Note, in this hypothetical example, that these clinical tests had not been used specifically for the problem of identifying the disease D. Thus preliminary analysis would be required to evaluate how well various combinations of clinical tests might work for this purpose.

Before the test results are analyzed for their reliabilities, it is instructive to examine how the tests will be used in predicting the disease, so that the data on the tests can be summarized in a form that can be used for that purpose. Specifically, suppose the physician is considering the use of four routine clinical tests, T_1, T_2, T_3, and T_4, and that the results of these tests on a particular patient are given by t_1, t_2, t_3, and t_4, respectively. The physician is interested in knowing what is the likelihood of disease D, given the results of the four tests. One way of expressing this mathematically is by a conditional probability: $\Pr(D|t_1, t_2, t_3, t_4)$, where "Pr" represents "probability" and "|" is read as "given that." The preceding conditional probability is thus read as: "the probability of disease D given that tests T_1, T_2, T_3, and T_4 gave the results t_1, t_2, t_3, and t_4, respectively." A way of computing this probability based on past information on the tests and the current test results is through the use of Bayes' theorem (or Bayes' formula). The prediction, $\Pr(D|t_1, t_2, t_3, t_4)$, based on Bayes' formula, which will hereafter be termed the "Bayesian prediction," is given as follows:

$$\Pr(D|t_1, t_2, t_3, t_4) = \frac{\Pr(D)\Pr(t_1, t_2, t_3, t_4|D)}{\Pr(D)\Pr(t_1, t_2, t_3, t_4|D) + \Pr(ND)\Pr(t_1, t_2, t_3, t_4|ND)}$$

Preliminary Analysis

where $\Pr(D)$ is the prior (before testing) probability of disease, $\Pr(ND)$ is the prior probability of no disease, and $\Pr(t_1, t_2, t_3, t_4|D)$ and $\Pr(t_1, t_2, t_3, t_4|ND)$ are the conditional probabilities of getting the given test results. This is the approach taken in the CPBS.

Note that in the preceding formula, the conditional probability, $\Pr(t_1, t_2, t_3, t_4|D)$, makes sense only if the test results are given as discrete probability densities (since otherwise these probabilities would be zero). Specifically, since we are trying to distinguish between *two states of nature*, namely, disease and no disease, we will assume that each of the test results be expressed as one of *two possible results*—either positive or negative for the disease. For example, $t_1, t_2, t_3,$ and t_4 could represent the results $+, +, -, +$, respectively, if T_1 was positive for the disease, T_2 was positive for the disease, T_3 was negative for the disease, and T_4 was positive for the disease.

We can estimate the probability of obtaining various combinations of test results in patients with the disease, $\Pr(t_1, t_2, t_3, t_4|D)$, if we have a data base that contains the results of all four tests on patients that have the disease D. For example, if the results for $T_1, T_2, T_3,$ and T_4 were $+, +, -, +$, respectively, and our past data on the tests showed that out of all the patients that had the disease, this sequence of results for the tests was obtained 10.0% of the time, then we would estimate $\Pr(+ + - +|D)$ to be 0.10. Note that if this estimate was based on results on only 10 patients, then our confidence in this estimate would be relatively low in comparison to one based on 100 patients. The probability $\Pr(t_1, t_2, t_3, t_4|ND)$ can be estimated in a similar fashion from data on patients that did not have the disease.

Alternatively, if we could establish that the four different tests were statistically independent when applied on patients that have the given disease, then the conditional probability $\Pr(t_1, t_2, t_3, t_4|D)$ could be simply computed as the product of the individual conditional probabilities for the four tests, namely, $\Pr(t_1|D)\Pr(t_2|D)\Pr(t_3|D)\Pr(t_4|D)$. We can see that each of these probabilities is simply a reflection of the past performance of the individual test on patients that had the disease. For example, in order to estimate $\Pr(t_1|D)$, where T_1 was positive for the disease, we would look at the data base and see what proportion of the diseased patients tested positive in T_1. The reliability of our estimate for $\Pr(t_1|D)$ would depend on how much data we had on T_1 for the diseased patients. A similar statement could be made for the conditional probability $\Pr(t_1, t_2, t_3, t_4|ND)$.

Sometimes the results of one test seem to correlate very strongly with the results of another test. For example, both an elevated temperature and an elevated white blood cell count are often present in patients with an infection. It is important to recognize these types of relationships among the tests because (1) if the correlations are strong enough, then only one of the tests would be needed since the other test would essentially provide redundant information, and (2) such correlations would imply that the tests were not statistically independent. If the correlations were not recognized, we might mistakenly use the preceding estimate for $\Pr(t_1, t_2, t_3, t_4 | D)$ (i.e., the estimate computed as the product of the individual performances of the tests), and this could produce a significant error in the Bayesian prediction if the correlations were not somehow accounted for.

The preceding example has been used as a vehicle to illustrate some of the key issues involved in preliminary analysis, namely, (1) the kind of data that we need in our data base (which is dictated by the purpose for which the data are going to be used) and (2) the important pieces of information that should be summarized from the data base (which include estimates for the reliabilities of the individual tests and estimates of how the tests would work in groups based on evaluating the interdependencies and correlations among the tests). This chapter provides several approaches for addressing these issues. In particular, in Section 3.1 we discuss the characteristics of the data base that are required by the CPBS methodology in order to estimate the reliabilities of the tests and the interdependencies among the tests. In Section 3.2, we discuss the analysis of the individual tests and describe how to estimate the two major summary measures of individual test performance used in Bayes' theorem. Sections 3.3 and 3.4 both deal with issues involving the group performances of tests. In Section 3.3 we describe how cluster analysis (see also Section 2.2) can be used to explore relationships among the tests and relationships among the objects in the data base, and why this information might be of use in both selecting a battery of tests and interpreting the results of a battery of tests. Finally, in Section 3.4 we discuss how failure to recognize interdependencies among tests can produce errors in our estimate of battery performance and in the Bayesian predictions; and we describe a set of dependence measures that can be used to test for the presence of dependencies and can be used as a measure of the strength of dependence in Bayes' formula.

3.1. DATA BASE CONSIDERATIONS

As indicated in the introduction to this chapter, preliminary analysis (if it is required) begins with the analysis of any data base containing test results on objects with *known properties*. A minimal requirement is that results of tests on objects known both to have and not to have the property of interest would need to be present in the data base in order to compute the conditional probabilities [e.g., $\Pr(t_1, t_2, t_3, t_4 | D)$ and

Preliminary Analysis

$Pr(t_1, t_2, t_3, t_4 | ND)$ in the preceding example] that are required in Bayes' formula.

In general, data bases containing test results can have a number of characteristics. For example, preliminary analysis requires that we identify in the data base at least some objects known to have and known not to have the particular property of interest. In trying to classify whether an object has or does not have the property of interest, we might have available to us some definitive set of evidence or a "gold standard" (e.g., in the case where the property of interest is the disease pneumonia, then a definitive test might be a blood culture; in the case where the property of interest is the carcinogenicity of a chemical to humans, then a definitive set of evidence might be results of epidemiological studies on humans); or we might have available to us a set of evidence that would allow us to infer whether the property is present or not (e.g., in trying to classify specific chemicals as carcinogenic to humans, since human testing is unacceptable, results from animal bioassays are used to infer whether a particular chemical is carcinogenic to humans or not).

In the former case, where a "gold standard" is available, each object that has been tested by the gold standard can be classified strictly as having or not having the property. In the latter case, where a set of evidence is used for inference, we may not be able to make such a strict classification. Instead we might be able to classify the objects by "strength of evidence." For example, in one of the data bases containing short-term test results on chemicals—the IARC (1982) data base—chemicals were classified as having sufficient evidence, limited evidence, or insufficient evidence to be called carcinogens.

Not all of the objects in a data base may have been tested by a gold standard or any other set of surrogate tests. In this case, the data base would contain objects having unknown properties. For example, in the 1985 Gene-Tox data base (which contains short-term test results on chemicals compiled by peer review of published literature; see Palajda and Rosenkranz, 1985) there were 465 known carcinogens, 84 known noncarcinogens, and 1650 chemicals of unknown carcinogenicity.

The test results in a data base can also take on a number of forms. The result of a test may be given as positive (+), negative (−), or equivocal (?) for the property of interest. For example, in the previously mentioned Gene-Tox data base, each test result on a particular chemical is summarized by either no data available, a negative response, a questionable response, or a positive response. The form of a test result might be given as a continuous-scaled value. For example, in the example problem involving disease diagnosis, tests such as white blood cell counts and oral or rectal temperatures take on such a form. If the form of the test was positive or negative, then the strengths of results might also be given. For example, genotoxicity can be expressed in terms of units of potency (Waters et al., 1986).

The data base may be "completely filled," i.e., every object tested by every test, or (more likely) it will contain many gaps. For example, in the Gene-Tox data base, the number of carcinogens tested by any one test ranged from 6 for one of the tests to 170 for the test that was used most often; and this from among over 400 known carcinogens. One of the objectives of the National Toxicology Program (NTP) in a recent effort to generate data on short-term tests was to use a single protocol and identical chemical batches for each of the assays in the data base (Tennant et al., 1987).

In addition to differences in the forms that the elements in a data base can take, the criterion used to decide which test results should be included in the data base may be different depending on the specific problem. These differences can be in terms of deciding both which tests should be included and which objects should be included. The determination of which tests should be included may be based on (1) the desire to include a "representative set of tests" (e.g., data bases consisting of representative genetic and genotoxic end points in prokaryotic and eukaryotic systems might be considered if batteries of these tests were going to be used to predict chemical carcinogenicity), or (2) (as in the Gene-Tox data base) the desire to include as many tests as possible whose test results have passed specific standards.

For those data bases compiled to include a "representative set" of objects or tests, the definition of what is meant by "representative set" can also vary depending on the purpose for which the data were compiled. "Representative set" might mean a random sample of the objects or tests (e.g., data are collected to reflect a random set of chemicals found in the environment, or to ensure that chemicals common in the environment and in our lifestyles are included in the appropriate frequency in the data base); or it may mean a structured set of objects or tests (e.g., data are collected to include examples from most of the major classes of organic chemicals, to include noncarcinogens with structures similar to carcinogens, or to include an equal number of carcinogens and noncarcinogens).

Preliminary analysis can begin with any of the preceding types of data bases. There is only one characteristic of the data base that must be present in order to evaluate the performance characteristics of the tests, namely, each test in the data base must have at least some results on objects both known to have the property and known not to have the property of interest. We noted in the introduction to this chapter that because the CPBS uses Bayes' theorem for prediction, we require that the tests be given as positive or negative for the property. If the results of a test are not given as positive or negative (e.g., in the case where the results are continuous-valued), then the test results can be translated into positive and negative results *before* proceeding with preliminary analysis. This would require determining the range of values over which the given test would indicate the presence of the property and the range that would indicate the absence of the property.

It is not necessary for all of the objects in the data base to have known properties. We will see that as long as we have a sufficient number of objects known to have and not have the property of interest, we will still be able to estimate the performance characteristics; and that data on objects with unknown properties (e.g., short-term test results on chemicals of unknown carcinogenic potential) can also be of some use. In particular, in the case where there are many gaps in the data base, cluster analysis on objects with unknown properties can sometimes help to infer the selective performances of some of the tests, as well as the relationships among the tests. It could, of course, happen that the data on some of the tests are so sparse that it is impossible to estimate the performance characteristics. In this case we would recommend that an alternate reliable source for the summary information (i.e., conditional probabilities used in Bayes' theorem and the performances and correlations among the tests) be found.

Ideally, we would like the data base to contain a large number of results for each test on a representative set of objects both with and without the property of interest; furthermore, we would like the data base to have no gaps. For example, if physicians were interested in predicting the presence or absence of a particular disease in patients of a particular age group, we would like test results on a large number of individuals in that particular age group that both have and do not have the disease. Similarly, if a chemical company is interested in using the CPBS for interpreting short-term test results on a new product under development, desirable properties for the data base would be that it would contain a sufficiently large number of test results (of the short-term tests they plan to use) on a representative set of chemicals that includes both known carcinogens and known non-carcinogens.

It is seldom possible to develop or retrieve an *ideal* data base with the above characteristics. Either the required information on the tests does not currently exist, or the cost of developing such a data base may be very high. If one would like to use the CPBS to its best advantage, an effort should be made to construct a data base as close to ideal as possible. One way to do this would be to collect as many existing data as possible and then to do selective testing in order to fill in the gaps in the data base.

Although an "ideal" data base is highly desirable, the CPBS can be used on *any* data base—even data bases with a large number of gaps. It should be noted, however, that the results of the CPBS are only as good as the underlying data. Thus, if the data base is far from complete and one decides to use the CPBS, the results of the CPBS should be viewed only as an additional source of information that can be used along with other sources of evidence to support decisions.

3.2. ANALYSIS OF TEST PERFORMANCE

Suppose that we have a choice of several tests that could be used in helping to identify whether a particular property is present in a specific object. However, we are not quite sure how well the tests would work or which one(s) to use. Our desire would be to find and utilize the tests that are the most *selective*, i.e., to find tests that would show positive responses whenever objects with the property were tested and would show negative responses whenever objects without the property were tested.

If we have a data base (as described in the preceding section) containing results of these tests on objects with known properties, one of the first things we might want to do is to evaluate how well these tests performed on the objects in the data base. There are a number of ways to evaluate and summarize the ability of a test to distinguish among objects that have the property of interest and objects that do not (see, for example, Greenes *et al.*, 1984; Galen and Gambino, 1975; Chankong *et al.*, 1985; and Pet-Edwards, 1986). The most common of these performance measures includes the *sensitivity* of a test (the ability of the test to identify objects that have the property of interest), the *specificity* of a test (the ability of the test to identify objects that do not have the property of interest), and the *accuracy* of a test (a combined measure of the abilities of the test to correctly identify objects both with and without the property of interest). Some other commonly used measures include the *predictivity* (Chankong *et al.*, 1985) or *predictive value* (Galen and Gambino, 1975), which is the strength of the Bayesian prediction (i.e., the magnitude of the conditional probability computed using Bayes' formula) that is obtained from a positive or negative test result; *assignment potential*, which is the probability that the Bayesian prediction probability will exceed a given threshold for management; and *assignment strength*, which is the average extent to which the Bayesian prediction probability would exceed the threshold for management.

Preliminary Analysis

To see which of the preceding performance measures might be most appropriate to use, let us examine Bayes' formula for the case where we have a single test result. In particular, suppose we are trying to estimate the likelihood that a property, P, is present, given that we obtained a positive result using test T_i. The Bayesian prediction can be given as follows:

$$\Pr(P|+_i) = \frac{\Pr(+_i|P)\Pr(P)}{\Pr(+_i|P)\Pr(P) + \Pr(+_i|NP)\Pr(NP)} \quad (3.2.1)$$

(In the preceding formula we have used "$+_i$" to represent the positive result given by T_i and "NP" to represent that the property is not present.)

We can see that in order to compute the Bayesian prediction based on the positive result of T_i, we need to have estimates for $\Pr(+_i|P)$ and $\Pr(+_i|NP)$. Note that $\Pr(+_i|P)$ is the likelihood that the test would give a positive result on an object with the property (i.e., the ability of the test to identify objects with the property). $\Pr(+_i|NP)$ can be computed as $[1 - \Pr(-_i|NP)]$, where "$-_i$" represents a negative result for T_i; and $\Pr(-_i|NP)$ is the likelihood that the test would give a negative result on an object where the property was not present. This latter probability would reflect the ability of the test to identify objects for which the property is not present. From the list of common performance indices given above, we see that the sensitivity and specificity would appear as natural components in Bayes' formula.

Because of their wide use and because they form natural components in Bayes' formula, we have employed the sensitivity and specificity to represent a composite measure of a test's *selective* ability. Mathematically, these two measures can be given as conditional probabilities as follows:

$\Pr(+_i|P)$: Probability that test i would give a positive result given that the test object has the property. This is the *sensitivity* of the test. (3.2.2)

$\Pr(-_i|NP)$: Probability that test i would give a negative result given that the test object does not have the property. This is the *specificity* of the test. (3.2.3)

From a computational standpoint, the two preceding probabilities (i.e., the sensitivity and specificity) can be estimated by the following two formulas for each test in the data base, respectively:

$$\alpha_i^+ = \frac{\text{Number of positive results of } T_i \text{ on objects with P}}{\text{Total number of objects with P tested}} \quad (3.2.4)$$

$$\alpha_i^- = \frac{\text{Number of negative results of } T_i \text{ on objects with NP}}{\text{Total number of objects with NP tested}} \quad (3.2.5)$$

The probabilities in Eqs. (3.2.2) and (3.2.3) are used to denote the theoretical sensitivity and specificity of test i and can be thought of as the limiting case when the number of objects within P and NP (P and NP are being referred to as populations of objects with and without the property) goes to infinity. Whenever we are referring to or when we are using estimates for the sensitivity and specificity [e.g., the ratios given in Eqs. (3.2.4) and (3.2.5) computed from the raw test scores], we will use the shorthand notation α_i^+ and α_i^-, respectively.

In the CPBS, we would like to obtain the best available estimates for the sensitivity and specificity of each test, since the reliability of these performance measures directly affects the reliability of the resulting Bayesian predictions. It is obvious that the number of objects tested will have a great impact on the reliability of the two estimates given in Eqs. (3.2.4) and (3.2.5). If an "ideal" data base containing a sufficiently large and representative set of objects and a large number of test results is used, then good estimates for the sensitivity and specificity of each test can be computed using Eqs. (3.2.4) and (3.2.5). However, when dealing with a data base that is less than ideal, the reliabilities of these two estimates could be affected.

In our experience, one of the most common characteristics that makes a data base less than ideal is sparseness (i.e., gaps in the data base). When a data base is sparse, the degrees of confidence associated with the two estimates can be very low and can differ for each of the tests depending on the number of objects that have been tested by each of the tests. An example of sparse data would be the short-term test results given in the Gene-Tox data base (the contents of the data base were published in Palajda and Rosenkranz, 1985). To obtain some idea of the reliability of the estimates for the sensitivity and specificity [given in Eqs. (3.2.4) and (3.2.5)] as a function of the number of objects tested, we can compute a confidence interval for each estimate. In particular, the following approximate $[100(1 - \beta)]\%$ confidence intervals can be computed for the two relative frequencies (or percentages) given in Eqs. (3.2.4) and (3.2.5).

Let Y/n represent the ratio given in Eq. (3.2.4) or Eq. (3.2.5); i.e., Y would represent either the number of positive results or negative results

Preliminary Analysis

(depending on whether the sensitivity or specificity is being computed), and n would represent the number of objects with P or NP tested. Note that under assumptions of independence and a constant probability p, Y is a random variable and has a binomial distribution $b(n, p)$, where p is the probability that the test gives a positive result on objects with P or a negative result on objects with NP (i.e., p is the theoretical sensitivity or specificity). Thus a binomial table can be used to compute the $[100(1 - \beta)]\%$ confidence intervals for the estimates in Eqs. (3.2.4) and (3.2.5). Namely, we find the minimum value of i (where i is an integer) that satisfies

$$\Pr([y - i] < Y < [y + i]) \geq 1 - \beta \tag{3.2.6}$$

(where β is the desired level of significance, and y is the sample value for the random variable Y) through the use of a binomial table with parameters n and $p = y/n$. And, after finding the value for i, then the approximate $[100(1 - \beta)]\%$ confidence interval would be given by

$$\frac{y - i}{n} < p^* < \frac{y + i}{n} \tag{3.2.7}$$

where p^* is the theoretical value for the sensitivity or specificity. Note that the ratio $(y + i)/n$ should not exceed the value one and the ratio $(y - i)/n$ should not fall below zero since they both represent ranges on a probability. In the case where Eq. (3.2.7) produces results that extend beyond the range from zero to one, the results should be truncated. For example, if $y = 0$, $n = 2$, and $i = 1$; then Eq. (3.2.7) would give $-0.5 < p^* < 0.5$. The negative lower bound should be truncated to zero, and the result given as $0.0 < p^* < 0.5$. The following example illustrates the computation of a confidence interval using Eqs. (3.2.6) and (3.2.7).

EXAMPLE 3.2.1

In the Gene-Tox data base compiled by Palajda and Rosenkranz (1985), the short-term test *Proteus mirabilis* (Prm) was used on nine known carcinogens. Prm gave positive results for eight out of these nine carcinogens and gave a negative result for one. Using Eq. (3.2.4) the estimate for the sensitivity of Prm is $\alpha^+ = 0.889$. The reliability of this estimate can be computed using Eqs. (3.2.6) and (3.2.7). In particular, suppose we would like a 99% confidence interval for the sensitivity estimate (i.e., we would like to be 99% confident that the computed interval contains the true sensitivity). For convenience, a portion of the binomial table corresponding to $n = 9$ and $p = 0.9$ (0.9 is being used as an approximation for 0.889) is reproduced below:

y	$\Pr(Y = y)$
0	0.0000
1	0.0000
2	0.0000
3	0.0001
4	0.0008
5	0.0074
6	0.0446
7	0.1722
8	0.3874
9	0.3874

Using the preceding table, we obtain the following results for Eq. (3.2.6):

i	$\Pr(8 - i < Y < 8 + i)$
0	0.0000
1	0.3874
2	0.9470
3	0.9916

We see that for $i = 3$, the preceding probability exceeds our confidence level of 0.99; thus, according to Eq. (3.2.7), with $i = 3$, $y = 8$, and $n = 9$, the 99% confidence interval is given by $0.55 < \Pr(+|P) < 1.0$.

When n (the number of objects tested) in the ratio Y/n is large, then the following expression involving the binomial random variable Y has an approximate standard normal distribution:

$$\frac{(Y/n - p)}{p(1 - p)/n} \sim N(0, 1)$$

Thus, for "sufficiently large" sample sizes [a rule often stated is that n is "sufficiently large" if $np \geq 5$ and $n(1 - p) \geq 5$], a simpler confidence interval can be computed based on the standard normal distribution. In this case the approximate $[100(1 - \beta)]\%$ confidence interval can be computed as

$$y/n - Z_{1-\beta/2}\left(\frac{y/n(1 - y/n)}{n}\right)^{1/2}$$

$$< p^* < y/n + Z_{1-\beta/2}\left(\frac{y/n(1 - y/n)}{n}\right)^{1/2} \quad (3.2.8)$$

Preliminary Analysis

where n is the number of objects tested, p^* is the true value for the sensitivity or specificity, y/n is the estimate for the sensitivity or specificity given by Eq. (3.2.4) or Eq. (3.2.5), respectively, β is the level of significance, and $Z_{1-\beta/2}$ is the standard normal variate. For example, for an approximate 95% confidence interval, $\beta = 0.05$, $1 - \beta/2 = 0.975$, and $Z_{1-\beta/2} = 1.96$; and for an approximate 90% confidence interval, $\beta = 0.1$ and $Z_{1-\beta/2} = 1.645$. The following example illustrates the use of Eq. (3.2.8) for computing a confidence interval:

EXAMPLE 3.2.2

In the Gene-Tox data base (Palajda and Rosenkranz, 1985), the short-term test *Salmonella* (Sty) was used on 170 known carcinogens. In 104 of the cases, Sty gave positive results. From Eq. (3.2.4), the estimate for the sensitivity of Sty is given by $\alpha^+ = 0.612$. Note that $n\alpha^+ = (170)(0.612) = 104$ and $n(1 - \alpha^+) = (170)(0.388) = 66$. Thus n is "sufficiently large" to warrant the use of the standard normal approximation given by Eq. (3.2.8). If we wish to compute a 95% confidence interval for the theoretical sensitivity of Sty, then $n = 170$, $y/n = 0.612$, and $Z_{1-\beta/2} = 1.96$. The confidence interval would thus be given as $0.539 < \Pr(+|P) < 0.685$.

When the reliabilities of our estimates for the sensitivities or specificities are not very high [i.e., the confidence intervals computed by either Eq. (3.2.7) or Eq. (3.2.8) are wide], then a heuristic approach can be used to evaluate whether the estimates, α_i^+ and α_i^-, reflect the "true" sensitivity and specificity of test i. In particular, we make the following observations: Tests that give positive and negative responses on the same set of objects should have the same sensitivities and specificities. Thus if we are able to group the tests that give similar responses based on an expanded data base, possibly including test results on objects having unknown properties (as well as known ones), then we should be able to deduce that tests within such a group would have the same sensitivity and/or specificity. If, in addition, the sensitivity and/or specificity of a test within the group is known with a high degree of confidence, then the estimates of the other tests within the group can be strengthened by this information (Pet-Edwards et al., 1985b).

Cluster analysis (see Chapter 2, Section 2.2; and Section 3.3 of this chapter for more details) can be used to help determine which of the tests are most similar to one another in terms of their responses. To apply cluster analysis, comparisons are made between the responses of each pair of tests to determine the similarity between pairs of tests. These pairwise similarities are utilized in several different hierarchical clustering schemes to uncover the natural groupings (clusters) of tests. Clusters that are formed in this manner would be a characteristic of the responses of the tests on the set of objects. The sensitivities and specificities are also characteristics of the responses of the tests. Thus tests within a cluster having a high degree of similarity should give similar responses on the same set of objects, and consequently, should have similar performance indices.

The preceding observations led us to the following scheme for estimating the values and reliabilities for the sensitivities and specificities of the tests in a data base:

PRELIMINARY ANALYSIS: CPBS PROCEDURE FOR ESTIMATING TEST PERFORMANCE

Step 1. Use Eqs. (3.2.4) and (3.2.5) to estimate the sensitivity and specificity of each test in the data base.

Step 2. Compute a confidence interval for each estimate produced in Step 1. Use Eqs. (3.2.6) and (3.2.7) for small sample sizes (say less than 10 objects tested) and Eq. (3.2.8) for larger sample sizes (say 10 or more objects tested).

Step 3. If all confidence intervals are reasonably small, stop here. Otherwise use cluster analysis on the data base to group the tests based on the similarities among their responses.

Step 4. Use the results of cluster analysis to "improve" the estimates for the sensitivities and specificities in the data base: for each cluster, select the test that was used on the most objects and assign the sensitivity or specificity of that test to each of the remaining tests in the cluster.

The following is an example of the CPBS procedure for preliminary analysis where the procedure was applied to the Gene-Tox data base (adapted from Pet-Edwards *et al.*, 1985a).

EXAMPLE 3.2.3

The short-term tests in the Gene-Tox data base (see the Appendix for a list of abbreviations) were analyzed using the preceding four-step procedure to see how well each test worked in identifying the known carcinogens and known noncarcinogens in the data base. Tables 3.2.1 and 3.2.2 summarize the results of this procedure on the Gene-Tox data base. The results for Step 1 of the procedure are given in column two, marked "direct calculation." An examination of the number of chemicals tested (see Palajda and Rosenkranz, 1985) indicated that in several cases there were insufficient

TABLE 3.2.1. Estimates for the Sensitivities of the Short-Term Assays in the Gene-Tox Data Base[a]

Assay	Direct calculation	Confidence interval[b]	After cluster analysis Improved estimate	Improved interval
plA	0.836	$0.747 \leq p \leq 0.925$*	0.836	$0.747 \leq p \leq 0.925$
Bsr	0.719	$0.563 \leq p \leq 0.875$*	0.906	$0.827 \leq p \leq 0.985$
Sty[c]	0.612	$0.539 \leq p \leq 0.685$*	0.612	$0.539 \leq p \leq 0.685$
PrM[c]	0.889	$0.550 \leq p \leq 1.00$	0.836	$0.747 \leq p \leq 0.925$
EcW[c]	0.633	$0.461 \leq p \leq 0.805$*	0.612	$0.539 \leq p \leq 0.685$
Mly	0.950	$0.854 \leq p \leq 1.00$	0.836	$0.747 \leq p \leq 0.925$
V79	0.781	$0.638 \leq p \leq 0.924$*	0.781	$0.638 \leq p \leq 0.924$
CHO	1.00	$0.700 \leq p \leq 1.00$*	0.781	$0.638 \leq p \leq 0.924$
DRL[c]	0.803	$0.703 \leq p \leq 0.903$*	0.836	$0.747 \leq p \leq 0.925$
DCM	0.826	$0.671 \leq p \leq 0.981$*	0.781	$0.638 \leq p \leq 0.924$
Msl	0.636	$0.352 \leq p \leq 0.920$*	0.333	$0.066 \leq p \leq 0.600$
Msp	0.857	$0.674 \leq p \leq 1.00$*	0.757	$0.659 \leq p \leq 0.855$
Cbm	0.818	$0.590 \leq p \leq 1.00$*	0.836	$0.747 \leq p \leq 0.925$
Csg	0.650	$0.250 \leq p \leq 0.950$	[d]	[d]
UDS[c]	0.560	$0.422 \leq p \leq 0.698$*	0.612	$0.539 \leq p \leq 0.685$
Drp	0.947	$0.729 \leq p \leq 1.00$*	0.890	$0.818 \leq p \leq 0.962$
Spo	0.882	$0.729 \leq p \leq 1.00$*	0.890	$0.818 \leq p \leq 0.962$
Csp	0.333	$0.066 \leq p \leq 0.600$*	0.333	$0.066 \leq p \leq 0.600$
Cle	0.955	$0.868 \leq p \leq 1.00$*	0.836	$0.747 \leq p \leq 0.925$
Mnt	0.929	$0.834 \leq p \leq 1.00$*	0.836	$0.747 \leq p \leq 0.925$
Htr	0.889	$0.550 \leq p \leq 1.00$	0.890	$0.818 \leq p \leq 0.962$
3T3	0.552	$0.371 \leq p \leq 0.733$*	0.781	$0.638 \leq p \leq 0.924$
SHE[c]	0.906	$0.827 \leq p \leq 0.985$*	0.906	$0.827 \leq p \leq 0.985$
BHK[c]	0.781	$0.638 \leq p \leq 0.924$*	0.906	$0.827 \leq p \leq 0.985$
HMA[c]	0.757	$0.659 \leq p \leq 0.855$*	0.757	$0.659 \leq p \leq 0.855$
Coo	0.714	$0.350 \leq p \leq 0.950$	0.757	$0.659 \leq p \leq 0.855$
Cvt[c]	0.820	$0.671 \leq p \leq 0.981$*	0.890	$0.818 \leq p \leq 0.962$
Dan	0.941	$0.829 \leq p \leq 1.00$*	0.890	$0.818 \leq p \leq 0.962$
Bfl	0.680	$0.497 \leq p \leq 0.863$*	0.757	$0.659 \leq p \leq 0.855$
SCE[c]	0.895	$0.789 \leq p \leq 0.992$*	0.890	$0.818 \leq p \leq 0.962$
VET[c]	0.890	$0.818 \leq p \leq 0.962$*	0.890	$0.818 \leq p \leq 0.962$
C3H	1.00	$1.000 \leq p \leq 1.00$*	0.890	$0.818 \leq p \leq 0.962$
MPR	0.833	$0.400 \leq p \leq 1.00$	0.890	$0.818 \leq p \leq 0.962$

[a] Taken from Pet-Edwards et al. (1985a).
[b] Intervals marked by an asterisk are at 95% confidence level, all others computed by binomial distribution.
[c] Tests selected for battery construction.
[d] No improvement.

TABLE 3.2.2. Estimates for the Specificities of the Short-Term Assays in the Gene-Tox Data Base[a]

Assay	Direct calculation	Confidence interval	After cluster analysis	
			Improved estimates	Improved interval
plA	0.00[b]	No information	[c]	[c]
Bsr	0.500[b]	No information	[c]	[c]
Sty[d]	0.806	$0.667 \leq p \leq 0.945$	0.806	$0.667 \leq p \leq 0.945$
PrM[d]	0.500	$0.100 \leq p \leq 1.00$	[c]	[c]
EcW[d]	0.857	$0.450 \leq p \leq 1.00$	0.806	$0.667 \leq p \leq 0.945$
Mly	[e]	No information	[c]	[c]
V79	0.00[b]	No information	[c]	[c]
CHO	[e]	No information	[c]	[c]
DRL[d]	0.800	$0.350 \leq p \leq 1.00$	0.806	$0.667 \leq p \leq 0.945$
DCM	[e]	No information	[c]	[c]
Msl	[e]	No information	[c]	[c]
Msp	[e]	No information	[c]	[c]
Cbm	[e]	No information	[c]	[c]
Csg	[e]	No information	[c]	[c]
UDS[d]	1.00	$0.600 \leq p \leq 1.00$	0.806	$0.667 \leq p \leq 0.945$
Drp	[e]	No information	[c]	[c]
Spo	[e]	No information	[c]	[c]
Csp	[e]	No information	[c]	[c]
Cle	[e]	No information	[c]	[c]
Mnt	[e]	No information	[c]	[c]
Htr	[e]	No information	[c]	[c]
3T3	0.00[b]	No information	[c]	[c]
SHE[d]	0.667	$0.100 \leq p \leq 1.00$	0.667	$0.100 \leq p \leq 1.00$
BHK[d]	0.800	$0.350 \leq p \leq 1.00$	0.800	$0.350 \leq p \leq 1.00$
HMA[d]	1.00	$0.200 \leq p \leq 1.00$	0.800	$0.350 \leq p \leq 1.00$
Coo	[e]	No information	[c]	[c]
Cvt[d]	0.00[b]	No information	0.667	$0.100 \leq p \leq 1.00$
Dan	0.00[b]	No information	[c]	[c]
Bfl	0.00[b]	No information	[c]	[c]
SCE[d]	0.00[b]	No information	0.667	$0.100 \leq p \leq 1.00$
VET[d]	0.444	$0.150 \leq p \leq 0.750$	[c]	[c]
C3H	0.00[b]	No information	[c]	[c]
MPR	[e]	No information	[c]	[c]

[a] Taken (corrected) from Pet-Edwards *et al.* (1985a).
[b] Not significant.
[c] No improvement.
[d] Tests selected for battery construction.
[e] No noncarcinogens tested.

numbers of carcinogens tested to give accurate estimates for the sensitivities (e.g., Csg and MPR only tested six known carcinogens), and the situation was even worse for the number of noncarcinogens tested. Many assays were not used on any known noncarcinogens, and only one assay tested more than 10 noncarcinogens, namely, Sty.

Step 2 of the procedure was used to examine how "good" the estimates were for the sensitivities and specificities. Confidence intervals at the 95% level were computed for each of the estimates and are displayed in column three of Tables 3.2.1 and 3.2.2. Each of the confidence intervals marked by an asterisk was computed using Eq. (3.2.8), and the remaining intervals (which were based on less than 10 samples) were computed using the binomial approximation [i.e., Eqs. (3.2.6) and (3.2.7)]. Notice that in many cases the data base did not provide information about the specificity of the assay, and that many of the confidence intervals were very wide.

Since some of the confidence intervals were wide, Step 3 (cluster analysis) was applied to an expanded data base containing both known carcinogens and noncarcinogens as well as chemicals of unknown carcinogenicity. [Note that Steps 1 and 2 use only test results on known carcinogens and noncarcinogens, whereas cluster analysis can be applied to chemicals of unknown carcinogenicity as well.] This expanded data base increased the number of chemicals tested (i.e., sample size) from 550 to over 2000. The results of the complete-link method (see Chapter 2 for a general discussion of the complete-link method and Section 3.3 for a detailed description of how these results were obtained from the Gene-Tox data base) are displayed in Tables 3.2.3 and 3.2.4. The clusters in Table 3.2.3 were obtained by applying the complete-link method to the similarity between the negative responses of the tests (see Example 3.3.1 in Section 3.3); the clusters in Table 3.2.4 were obtained by applying the complete-link

TABLE 3.2.3. Clusters Obtained by the Complete-Link Method; Similarities between Negative Responses Are Employed[a,b]

No.	Cluster
1	UDS, DRL, Sty,* EcW
2	V79, 3T3*
3	SCE, SHE,* Cvt
4	BHK,* HMA

[a] Taken from Pet-Edwards et al. (1985b).
[b] The asterisk indicates the assay within the cluster that tested the most chemicals.

TABLE 3.2.4. Clusters Obtained by the
Complete-Link Method; Simple Matching
Coefficients Are Employed[a,b]

No.	Cluster
1	Msl, Csp*
2	3T3, DCM, V79,* CHO
3	UDS, Sty,* EcW
4	Htr, Cvt, MPR, Spo, SCE, Drp, Dan, VET,* C3H
5	Coo, Bfl, Msp, HMA*
6	Bsr, SHE,* BHK
7	Cle, plA,* PrM, Mly, DRL, Cbm, Mnt

[a] Taken from Pet-Edwards et al. (1985b).
[b] The asterisk indicates the assay within the cluster that tested the most chemicals.

method to simple matching coefficients and are used to "improve" the sensitivities.

Step 4 of the procedure was used to try to improve some of the estimates for the sensitivities and specificities. The asterisk within each of the clusters in Tables 3.2.3 and 3.2.4 indicates which assay within each cluster tested the most chemicals. The sensitivities and confidence intervals of the assays marked with asterisks in Table 3.2.4 were assigned to each of the remaining tests within the clusters (see columns 4 and 5 in Table 3.2.1), and the specificities and confidence intervals of the assays marked with asterisks in Table 3.2.3 were assigned to each of the remaining tests in the clusters (see columns 4 and 5 in Table 3.2.2).

From an examination of Tables 3.2.1 and 3.2.2 in this example, we can see that the procedure for improving test performance estimates using cluster analysis helped to improve the estimates of the sensitivities and, on a limited basis, the specificities in the Gene-Tox data base in two ways: (1) in a few cases, the cluster analysis results were able to provide us with estimates for the specificities of assays that had not been applied to any known noncarcinogens (see for example, Cvt and SCE in Table 3.2.2), and (2) the results of cluster analysis improved the estimates for some of the sensitivities and specificities by providing more confidence (i.e., by reducing the size of the confidence intervals) in the computed estimates (see for example, DRL and HMA in Table 3.2.2 and UDS, Bfl, and SCE in Table 3.2.1). It should be noted, however, that although this procedure did help to improve a few of the estimates, there were still a fairly large number of tests for which there were no estimates available as well as some with very wide confidence intervals. The approach could not completely overcome the major deficiencies in this data base, namely, the large number of gaps and the small

Preliminary Analysis   85

number of objects without the property of interest (i.e., noncarcinogens) tested.

The preceding example demonstrates the approach used in the CPBS for analyzing the performances of the individual tests in a data base. It also clearly illustrates that no matter how good the analyses are, if the data are incomplete you would have no choice but to either collect more data or utilize the results "as is." If one chooses the latter option, this should only be done with great care and the results should only be considered as supportive evidence to be used in addition to other (possibly more conclusive) evidence. In any case, the results of this preliminary analysis of the individual tests would provide estimates for the performances of (at least some of) the individual tests for use in Bayesian prediction and battery selection. The analysis would also indicate which of the estimates are reliable and would indicate which of them are not. Since we plan to use these estimates in the process of interpreting specific battery results and in selecting batteries of tests, this information on the reliability of the performance estimates is clearly of importance in ensuring that the Bayesian predictions will be reliable.

3.3. EXPLORATORY ANALYSIS OF THE DATA

In the preceding section we described how the performances of individual tests in a data base can be evaluated and summarized. We will now begin to discuss how the group behavior of the tests can be analyzed and summarized. In order to evaluate the potential abilities of groups of tests to separate objects with the property of interest from those without, we can start by exploring the relationships that exist among the tests in the data base. In Chapter 2 we described an exploratory data analysis methodology called cluster analysis which is designed to uncover natural groupings of items (tests or objects) in a data base. In Section 3.2 we also talked about cluster analysis and said that it can sometimes be useful for (1) providing estimates for the sensitivity and/or specificity of a test when there is little or no information available to calculate these measures directly, and (2) enhancing our confidence in the estimates for the sensitivity and/or specificity computed from the data base. We also noted that cluster analysis can provide a means for extracting useful information from test results on objects with unknown properties in addition to the objects with known properties, thereby expanding the data base.

In addition to providing information useful for improving the estimates for test performance, cluster analysis can provide information that is useful for evaluating the group performances of the tests. We will see that it can provide indications of positive associations or dependencies among tests. This can be useful in battery selection since two tests that give practically the same results on a set of objects (which would, consequently, be positively dependent) would work no better than either one alone. Thus, the results of cluster analysis can indicate which tests can be substituted for one another in a battery that seeks to minimize time and/or cost.

Cluster analysis may also be used to help uncover natural groupings of the objects in the data base. Up to now, we have only described the potential usefulness of cluster analysis for grouping the tests. If instead we apply cluster analysis for grouping the objects, the results can be used to obtain an indication of whether the results of a group of tests are sufficient to separate objects that have the property of interest from those that do not. For example, if cluster analysis was applied to the chemicals in a data base containing short-term test results for carcinogenicity, and if we obtained two strong clusters where one consisted mostly of carcinogens and the other mostly of noncarcinogens, then we can be fairly confident that the tests can be used to separate carcinogens from noncarcinogens. There is, however, a difficulty in applying cluster analysis on the objects in a data base—namely, the size of the proximity matrix constructed from the data base (see Chapter 2 for a discussion of proximity matrices). The number of objects in a data base can be very large. In fact, the more objects in the data base, the better the data base is. Consequently, the proximity matrix constructed to summarize the similarities between all pairs of objects in the data base can be very large. For example, in the Gene-Tox data base there were approximately 2100 chemicals (objects) tested. If we would like to examine the similarities among the chemicals, then the resulting proximity matrix would have a dimension of 2100 × 2100. At the present, we are not aware of a commercially available computer package for cluster analysis that can handle a matrix of this size. Since in most instances the analysis of the similarities among the objects would be impractical to do, we will only discuss and illustrate the application of cluster analysis for grouping the tests in a data base. It should be noted, however, that the analysis of the similarities among the objects in a data base would proceed in an analogous fashion.

As discussed earlier, the data base utilized in the CPBS consists of positive and negative test results on a set of objects, and the objects may have known or unknown properties. In order to apply cluster analysis to such a data base, the data must first be transformed into a proximity matrix (see Section 2.2). To construct a proximity matrix from the test results, an index of proximity or "alikeness" must be established between all pairs of tests. In the CPBS, we utilize "simple matching coefficients" as a measure of the proximity (i.e., similarity) between the responses of each pair of tests:

Simple Matching Coefficients

$$S_{ij} = \frac{\text{Number of matching results between } T_i \text{ and } T_j}{\text{Number of objects tested by both } T_i \text{ and } T_j} \quad (3.3.1)$$

The simple matching coefficient given in Eq. (3.3.1) reflects the amount of information overlap between the pair of tests.

We noted earlier that cluster analysis can provide an indication of positive dependencies between tests. In particular, if the similarity between the responses of two tests is high enough, then the two tests will be positively dependent. In order to see why a "relatively high" value for S_{ij} between a given pair of tests T_i and T_j would indicate that the tests are positively dependent, consider the following illustration: Suppose we have two tests that have been applied to a randomly selected set of 100 objects. Suppose that the first test gave 80 positive results [i.e., $\Pr(+_1) = 0.80$] and the second test gave 70 positive results [i.e., $\Pr(+_2) = 0.70$]. If the two tests were statistically independent, then the probability that both tests will give positive results would be given as the product of the two individual probabilities, i.e., $\Pr(+_1, +_2) = 0.56$. A value for this joint probability that exceeds 0.56 would indicate that the two tests were positively dependent. In a similar manner, the joint probability of getting two negative results would be 0.06 if the two tests were statistically independent. Thus, if the two tests were statistically independent we would expect to get matching results approximately 62% of the time; i.e., $S_{12} = 0.62$ if T_1 and T_2 are independent. Consequently, a relatively high value for S_{12} (e.g., a value exceeding 0.62) would indicate that these two tests are positively dependent. A similar argument can be made for any two tests.

If the data base has many gaps (i.e., each object is not tested by all of the tests), then the reliabilities of the entries in the proximity matrix would differ depending on the values in the denominator of Eq. (3.3.1). Normally, cluster analysis should not be applied to such data without taking into account the different reliabilities of the similarity measures. There are currently no standard procedures for doing this. In the CPBS, we use a simple procedure to ensure that only the more reliable data are used in constructing the clusters. Namely, we construct a *complementarity matrix* containing the number of objects tested by each pair of tests [i.e., the values in the denominator of Eq. (3.3.1)] to go along with the proximity matrix. After the proximity matrix and its corresponding complementarity matrix have been constructed from the data, then the complete-link, single-link, and spanning-tree methods are applied in the manner described in Chapter 2 with the following modification: if a value in the complementarity matrix corresponding to a pair of tests is small (say less than 10), then the corresponding entry in the proximity matrix will be ignored in the construction of clusters. In this manner, proximity information that is based on small sample sizes will neither cause nor prevent two clusters from merging. The reader should refer to Section 2.2 for the specific details of the three cluster analysis methodologies.

The results of the three hierarchical clustering methods are examined in order to ascertain which tests form groups. The "end products" of these examinations are three lists of groups (clusters) of tests—one list corresponding to each of the three methods. The analyst may then want to merge the three lists, so that only the "strongest" groups are listed. Part of this examination involves a close scrutiny of the clusters and a comparison of the results with those that would be expected based on prior knowledge about the tests. This examination process is part of the process of cluster validation. Here creativity, experience, and insight play an important role in trying to justify and understand why certain tests have given similar responses and consequently have formed clusters.

The following is a summary of the steps used by the CPBS for exploratory analysis of the data.

PRELIMINARY ANALYSIS: CPBS PROCEDURE FOR EXPLORATORY DATA ANALYSIS

Step 1. Construct a proximity matrix corresponding to the measure of proximity given in Eq. (3.3.1) and compute the complementarity matrix based on the denominators of the proximity measures.

Step 2. Apply the complete-link, single-link, and spanning tree methods (each with the previously described modification for unreliable data) to the proximity-complementarity matrix pair.

Step 3. Validate and explore the results, and derive lists of clusters for the three cluster methodologies.

The following example [adapted from Pet-Edwards *et al.* (1985b)] illustrates a three-step procedure for cluster analysis as it was applied to the Gene-Tox data base.

EXAMPLE 3.3.1

The Gene-Tox data base compiled by Palajda and Rosenkranz (1985) consists of short-term test results on 549 chemicals of known carcinogenicity (i.e., chemicals for which animal carcinogenicity data are available) and approximately 1500 chemicals of unknown carcinogenicity. Cluster analysis

Preliminary Analysis

was applied to the data base in order to get a better understanding of the relationships (similarities) among the tests.

Step 1 was applied to the Gene-Tox data base. The proximity matrix (given in percent similarity) corresponding to Eq. (3.3.1) is given in Fig. 3.3.1 and the corresponding complementarity matrix (i.e., matrix containing the number of chemicals tested by each pair of tests) is given in Fig. 3.3.2. In this application, we also computed another measure of similarity, in the hope that it would provide information about the negative responses of the tests. This measure of proximity was defined as follows.

Negative Proximities

$$S_{ij} = \frac{\text{Number of results where both } T_i \text{ and } T_j \text{ were negative}}{\text{Number of results where either } T_i \text{ or } T_j \text{ was negative}}$$

If we can be assured that the number of false negatives among the negative responses is not too large, then the results of cluster analysis using the negative proximities can be used to enhance the estimates for the specificities in the data base (see Section 3.2).

The proximity and complementarity matrices corresponding to the negative proximity measures are given in Figs. 3.3.3 and 3.3.4, respectively.

Step 2 was applied to the two proximity-complementarity matrix pairs. Namely, the single-link, complete-link, and spanning tree methods were applied to the proximity matrices given in Figs. 3.3.1 and 3.3.3. These results are given in Figs. 3.3.5–3.3.10, respectively. It should be noted that entries corresponding to fewer than 8 chemicals in Fig. 3.3.2 were ignored for the proximity matrix given in Fig. 3.3.1, and entries corresponding to fewer than 5 chemicals in Fig. 3.3.4 were ignored for the proximity matrix given in Fig. 3.3.3.

Step 3 was then applied to validate and interpret the results of clustering methods and to derive two lists of clusters. From an examination of the single-link method results given in Fig. 3.3.5, we do not obtain a very clear picture of the groupings of the tests; however, for the most part, the results are consistent with the complete-link results given in Fig. 3.3.6. Focusing on a similarity level of 75%, the groups of tests given in Table 3.3.1 were found by the complete-link method for simple matching coefficients.

A similar examination of the cluster analysis results on the negative proximities (i.e., Figs. 3.3.8–3.3.10) indicates that the results for the single-link and complete-link methods were very similar, and that when a similarity level of 55% was considered, the complete-link clusters given in Table 3.3.2 were constructed.

	CAC	CHH	PLA	BSR	STY	PRM	ECW	MLY	V79	CHO	DRL	DCM	MSL	MSP	CBM	CSG	UDS	DRP	SPO	CSP	CLE	MNT	HTR	3T3	SHE	BHK	HMA	COO	CVT	DAN	BFL	SCE	VET	C3H	MPR
CAC	100	96	83	70	63	76	66	95	75	100	80	82	63	85	81	66	60	94	88	33	95	92	88	51	89	78	78	71	76	88	65	82	84	91	83
CHH		100	96	80	52	0	30	100	50	---	70	100	100	50	100	100	33	100	50	50	100	100	100	0	100	71	80	50	100	100	60	50	93	100	---
PLA			100	84	78	100	73	95	81	100	80	68	100	83	87	75	56	84	88	33	100	87	62	62	76	68	76	100	86	88	66	85	79	71	83
BSR				100	70	100	85	100	85	100	71	66	100	100	50	50	55	75	100	25	83	83	50	75	90	80	56	100	88	100	66	66	83	100	100
STY					100	81	90	79	77	92	84	66	54	64	63	40	79	76	86	38	66	69	70	62	78	80	69	50	70	68	65	73	68	76	75
PRM						100	100	100	66	100	57	75	100	100	50	0	100	100	80	0	100	75	66	100	80	100	80	0	50	75	100	80	100	100	100
ECW							100	90	81	100	83	75	100	100	50	66	81	83	100	50	66	69	87	66	78	88	80	100	85	75	66	66	85	75	100
MLY								100	88	100	100	71	66	90	100	100	78	100	100	42	83	100	100	55	84	80	88	100	81	100	66	94	92	100	100
V79									100	100	85	71	60	80	100	100	76	88	100	33	70	87	75	82	86	75	70	100	66	80	50	75	85	75	80
CHO										100	100	88	100	100	100	---	88	100	100	20	100	100	100	100	100	100	100	100	100	100	33	100	100	100	100
DRL											100	82	66	75	100	75	79	90	93	30	94	100	91	53	86	78	76	85	82	81	68	85	81	87	83
DCM												100	71	66	83	75	81	76	90	50	81	83	88	100	62	100	70	100	83	78	50	77	54	0	---
MSL													100	88	83	100	66	60	66	75	71	83	100	0	57	66	70	100	40	75	50	54	55	33	0
MSP														100	100	100	87	85	75	57	85	100	100	60	85	75	93	80	66	100	83	90	83	100	100
CBM															100	66	85	100	100	63	92	100	83	33	100	62	83	100	77	100	100	90	50	80	100
CSG																100	100	100	---	85	83	100	100	0	0	66	83	100	100	100	100	100	100	100	100
UDS																	100	94	90	28	75	66	100	68	66	80	68	80	72	88	53	77	50	62	88
DRP																		100	85	53	87	100	100	100	81	81	93	66	88	100	60	100	100	100	66
SPO																			100	20	57	100	100	100	100	100	80	100	100	83	40	100	100	100	100
CSP																				100	60	55	44	0	40	50	53	75	33	50	33	41	37	33	0
CLE																					100	100	85	50	80	62	87	100	92	85	66	82	62	100	100
MNT																						100	87	44	92	73	88	100	88	90	76	90	89	71	100
HTR																							100	0	100	100	80	100	83	71	75	90	80	100	---
3T3																								100	66	56	60	0	70	66	44	76	52	87	71
SHE																									100	96	72	66	78	85	73	87	80	87	63
BHK																										100	88	100	69	62	90	89	72	100	100
HMA																											100	75	70	78	81	92	69	83	87
COO																												100	100	83	33	100	66	100	---
CVT																													100	100	66	92	80	81	60
DAN																														100	50	88	90	66	100
BFL																															100	75	72	60	50
SCE																																100	87	88	100
VET																																	100	92	100
C3H																																		100	100
MPR																																			100

FIGURE 3.3.1. Percent similarity between pairs of assays using simple matching coefficients on the Gene-Tox data base (taken from Pet-Edwards *et al.*, 1985b).

Preliminary Analysis

FIGURE 3.3.2. Number of chemicals tested by each pair of tests in the Gene-Tox data base (taken from Pet-Edwards et al., 1985b).

	CAC	CHH	PLA	BSR	STY	PRM	ECW	MLY	V79	CHO	DRL	DCM	MSL	MSP	CBM	CSG	UDS	DRP	SPO	CSP	CLE	MNT	HTR	3T3	SHE	BHK	HMA	COO	CVT	DAN	BFL	SCE	VET	C3H	MPR
CAC	100	0	0	9	25	40	33	0	0	---	23	0	0	0	0	0	21	0	0	0	0	0	0	0	25	33	10	0	0	0	0	0	23	0	0
CHH		100	0	0	0	0	0	---	0	---	0	---	---	0	---	---	0	---	0	0	---	---	---	0	---	0	0	0	---	---	0	0	0	---	---
PLA			100	20	42	100	40	50	37	---	28	0	0	0	0	50	31	0	0	0	---	0	0	35	42	18	41	---	50	33	12	28	21	33	66
BSR				100	48	100	68	---	50	---	20	0	---	---	25	0	27	0	---	0	50	0	0	0	0	0	22	100	50	---	0	0	50	---	---
STY					100	57	81	16	53	50	69	12	28	0	20	0	67	0	40	20	12	9	25	39	48	57	36	0	35	0	13	33	27	20	20
PRM						100	100	---	0	---	0	0	---	---	0	0	100	---	0	0	100	50	0	---	66	100	50	0	33	0	---	50	100	---	---
ECW							100	0	33	---	77	0	100	---	0	0	75	0	---	20	25	0	66	0	40	66	50	---	0	0	0	0	60	0	---
MLY								100	0	---	---	0	0	0	---	---	25	---	---	0	0	---	---	0	33	0	33	---	0	---	0	50	0	---	---
V79									100	---	0	0	33	0	---	---	42	25	100	20	0	0	50	62	50	25	25	---	16	33	0	12	55	40	44
CHO										100	---	0	---	---	---	---	0	---	---	0	---	---	---	---	---	---	---	---	---	---	0	---	---	---	---
DRL											100	25	20	0	---	0	68	0	0	0	0	---	50	0	25	0	15	0	25	0	25	0	25	0	0
DCM												100	33	0	50	50	33	0	0	28	60	0	50	100	0	160	12	100	50	20	33	0	0	0	---
MSL													100	50	0	---	0	0	0	50	0	0	100	0	25	33	0	---	35	0	0	0	0	0	0
MSP														100	---	---	0	0	0	0	0	---	---	0	0	0	66	0	0	---	0	0	0	---	---
CBM															100	33	66	---	---	42	50	100	0	0	---	25	60	---	0	---	---	66	0	0	---
CSG																100	---	---	---	66	50	100	---	0	0	0	66	100	100	100	---	100	---	---	---
UDS																	100	0	50	0	40	0	100	42	41	55	42	0	37	50	25	12	11	0	66
DRP																		100	0	0	0	---	---	---	0	0	0	0	0	---	0	---	---	---	0
SPO																			100	0	0	---	---	---	---	---	0	---	---	0	25	---	100	---	---
CSP																				100	33	20	16	0	25	33	53	50	14	28	20	22	16	0	0
CLE																					100	---	0	0	0	25	50	100	50	50	0	25	0	---	---
MNT																						100	0	0	50	0	25	---	50	0	0	50	33	0	---
HTR																							100	0	100	100	0	---	0	0	50	0	0	100	---
3T3																								100	43	30	35	0	50	0	28	33	16	50	33
SHE																									100	85	40	0	55	50	20	55	40	50	11
BHK																										100	58	---	20	0	50	60	20	---	---
HMA																											100	0	36	33	33	57	13	0	66
COO																												100	100	50	0	---	0	---	---
CVT																													100	100	33	60	50	0	33
DAN																														100	0	0	0	0	---
BFL																															100	25	0	0	0
SCE																																100	54	33	---
VET																																	100	50	---
C3H																																		100	---
MPR																																			100

FIGURE 3.3.3. Percent similarity between pairs of assays using negative proximities on the Gene-Tox data base (taken from Pet-Edwards *et al.*, 1985b).

Preliminary Analysis

FIGURE 3.4. Number of chemicals tested by each pair of tests, where only results containing negative responses were counted (taken from Pet-Edwards et al., 1985b).

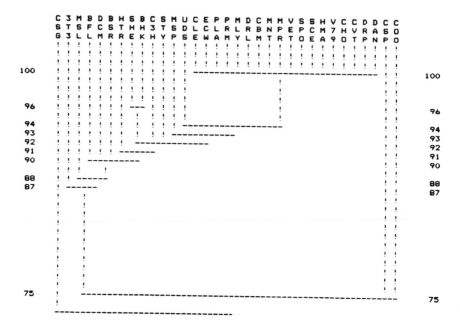

FIGURE 3.3.5. Single-link results on data given in Figs. 3.3.1 and 3.3.2 (taken from Pet-Edwards *et al.*, 1985b).

The spanning tree results give a very clear picture of which test(s) are most similar to a given test, but we cannot make inferences about how similar other (unlinked) pairs of tests are to each other. For example, in Fig. 3.3.7 we can see that CHO, Cbm, Htr, and Mly are each very similar to DRL. But we cannot say how similar, for example, CHO and Cbm are.

To illustrate how the results of cluster analysis may be validated by previous knowledge on the mechanistic basis of the various tests and how unexpected relationships can arise, the results given in Table 3.3.1 have been examined. An examination of the seven clusters listed in Table 3.3.1 reveals the following:

a. Cluster No. 1 is composed of two whole animal assays both of which involve germinal cells, i.e., the mouse specific locus assay and the *in vivo* spermatocyte cytogenetic assay.

b. Cluster No. 5 also is composed of whole animal assays involving primarily somatic mutations (Msp) or host-mediated assays using indicator systems (Bfl and HMA). On the other hand, the *in vivo* oocyte cytogenetics assay (Coo) would have been expected to be included in Cluster No. 1. Its inclusion in Cluster No. 5 might have mechanistic significance.

c. Cluster No. 7 is a mixed bag. It includes two bacterial DNA repair assays (plA and PrM) and three *in vivo* cytogenetic assays (Mnt, Cbm, Cle).

Preliminary Analysis

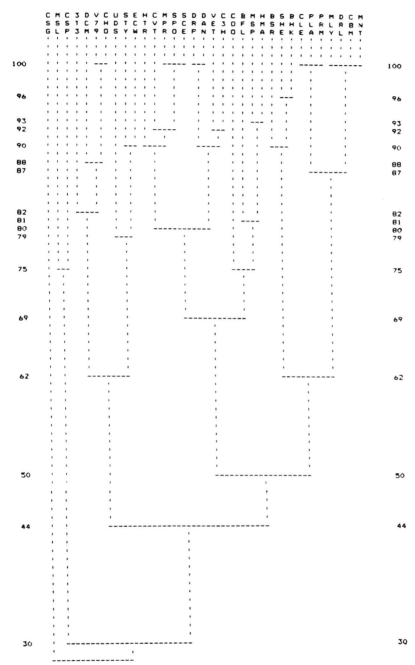

FIGURE 3.3.6. Complete-link results based on data given in Figs. 3.3.1 and 3.3.2 (taken from Pet-Edwards *et al.*, 1985b).

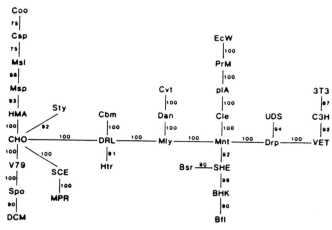

FIGURE 3.3.7. Spanning tree results based on data given in Figs. 3.3.1 and 3.3.2 (taken from Pet-Edwards et al., 1985b).

The presence of an *in vitro* murine gene mutation assay (Mly) is unexpected, as it might have been expected to cluster with the other gene mutation assays (V79 and CHO), although the latter were obtained in hamster cells. This cluster also contains DRL, the Drosophila sex-linked recessive lethal test. The significance of this is unknown (however, see below).

  d. The assays using Drosophila—DCM, DRL, and Dan—are in separate clusters. This suggests that for these assays the species is not as important as the genetic end point under investigation.

  e. Cluster No. 2 contains the two gene mutation assays (V79 and CHO) using hamster cells. The presence of the chromosomal mutation assay in Drosophila (DCM) cannot be explained, nor do we have an explanation for the presence of the murine transformation assay (3T3) in this group

TABLE 3.3.1. Clusters Obtained at the 75% Similarity Level Derived from Fig. 3.3.6[a]

No.	Cluster
1	Msl, Csp
2	3T3, DCM, V79, CHO
3	UDS, Sty, EcW
4	Htr, Cvt, MPR, Spo, SCE, Drp, Dan, VET, C3H
5	Coo, Bfl, Msp, HMA
6	Bsr, SHE, BHK
7	Cle, plA, PrM, Mly, DRL, Cbm, Mnt

[a] Taken from Pet-Edwards et al. (1985b).

Preliminary Analysis

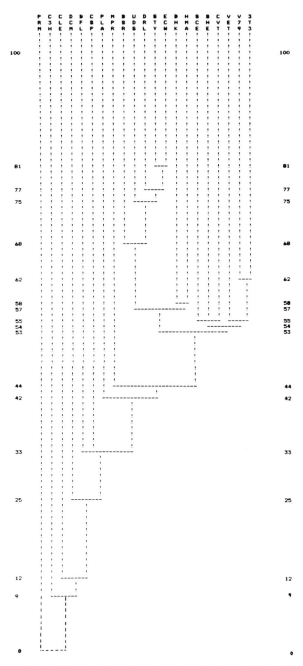

FIGURE 3.3.8. Single-link results based on data given in Figs. 3.3.3 and 3.3.4 (taken from Pet-Edwards *et al.*, 1985b).

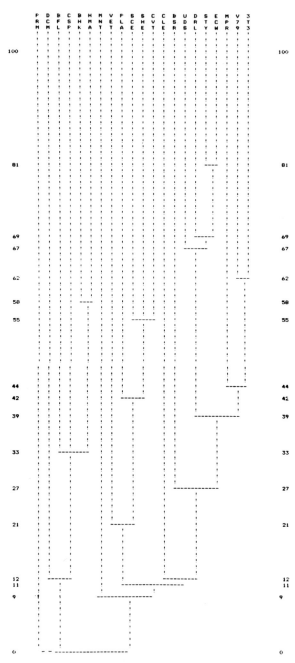

FIGURE 3.3.9. Complete-link results based on data given in Figs. 3.3.3 and 3.3.4 (taken from Pet-Edwards *et al.*, 1985b).

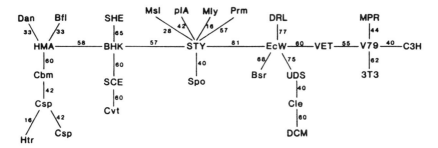

FIGURE 3.3.10. Spanning tree results based on data given in Figs. 3.3.3 and 3.3.4 (taken from Pet-Edwards *et al.*, 1985b).

rather than in cluster No. 4, which contains other murine transformation assays.

f. Cluster No. 3 contains the two bacterial mutagenicity assays (Sty and EcW) as well as a mammalian assay (UDS). It would have been expected, on the basis of current knowledge, that UDS would have clustered with SCE and DRP (cluster No. 4), since they all reflect damage to eukaryotic DNA.

g. Cluster No. 4 contains a number of murine transformation assays (C3H, VET, MPR); the association of SCE and DRP has already been mentioned. The significance of the inclusion of the other assays cannot be explained.

h. Cluster No. 6 contains two hamster transformation assays (BHK and SHE). The presence of Bsr, a bacterial DNA repair assay, is surprising as it would be expected to be associated with the two other bacterial repair assays (PrM and plA, cluster No. 7). It should be mentioned, however, that while Bsr, plA, and PrM measure the same phenomena, the permeability properties of *B. subtilis* (Bsr), a gram-positive microorganism, are vastly different from those of the gram-negative Proteus (PrM) or *E. coli* (plA).

TABLE 3.3.2. Clusters Obtained at the 55% Similarity Level Derived from Fig. 3.3.9[a]

No.	Cluster
1	UDS, DRL, Sty, EcW
2	V79, 3T3
3	SCE, SHE, Cvt
4	BHK, HMA

[a] Taken from Pet-Edwards *et al.* (1985b).

[It should be noted also that the Gene-Tox data base is not ideal. Thus the associations indicated above and the clusters obtained based on negative proximities may be a result of gaps in the data base, as well as false positives and false negatives in the data base. The significance of the associations will come under more rigorous scrutiny as additional data bases (e.g., the National Toxicology Program, Zeiger, 1982) are analyzed.]

We indicated earlier that the list of clusters obtained after applying cluster analysis to a data base can be utilized to improve the estimates for the sensitivities and specificities of tests and to provide an indication of positive statistical dependence between pairs of tests. The preceding example clearly indicates that cluster analysis can also be used simply as an exploratory tool to generate hypotheses about the way in which tests behave. In this respect, cluster analysis will sometimes uncover relationships that are not obvious from prior knowledge about the tests. Relationships that cannot be explained can initiate new ideas and can suggest new directions for research.

The use of the results from cluster analysis for improving the sensitivities and specificities of the tests was discussed and demonstrated in Section 3.2. The use of the clusters as an indicator of positive dependence, as well as other approaches to measuring and quantifying dependencies between test pairs, will be discussed and demonstrated in detail in Section 3.4. We will find that knowledge about the dependencies among the tests will be useful in both battery selection and prediction. How this information is used in battery selection and prediction will be discussed in Chapters 4 and 5, respectively.

3.4. TESTING FOR DEPENDENCIES AMONG TESTS

We have mentioned on several occasions that statistical dependence among tests can play an important role in the interpretation of multiple test results. We have shown that the results of cluster analysis can provide an indication of positive dependence among tests. In order to provide the reader with a better understanding of (1) the manner in which various types of dependencies can affect Bayesian predictions and (2) the extent to which dependencies can affect Bayesian predictions as a function of the number of tests and the sensitivities and specificities of the tests, we will begin this section with some examples illustrating how dependencies among tests affect the predictions based on Bayes' formula. We will then describe and illustrate the chi-square test for statistical dependence and will develop and describe

Preliminary Analysis

a new measure for dependence—the K_p measure—that we have utilized in the CPBS to account for dependencies in Bayes' formula.

3.4.1. The Effect of Statistical Dependence on Bayesian Predictions

We begin the discussion on the effect of statistical dependence with a simple example to illustrate how various "types of dependence" can affect the Bayesian prediction based on two positive test results. (The expression "types of dependence" will become clear as we proceed through the example.) Suppose we have two dichotomous tests, T_1 and T_2, both of which could be used to predict the presence or absence of a particular property. We will denote the presence of the property by "P" and the absence of the property by "NP" (not present). If both of the tests give positive results on a particular object, then the probability that the property is present in the object can be given by Bayes' formula:

$$\Pr(P|+_1,+_2) = \frac{\Pr(P)\Pr(+_1,+_2|P)}{\Pr(P)\Pr(+_1,+_2|P) + \Pr(NP)\Pr(+_1,+_2|NP)} \quad (3.4.1)$$

It will be more instructive for us to use an equivalent form of Bayes' formula given by dividing the numerator and denominator of the right-hand side of Eq. (3.4.1) by the expression $\Pr(P)\Pr(+_1,+_2|P)$. This gives us the following expression:

$$\Pr(P|+_1,+_2) = \frac{1}{1 + \dfrac{\Pr(NP)\Pr(+_1,+_2|NP)}{\Pr(P)\Pr(+_1,+_2|P)}} \quad (3.4.2)$$

From an examination of Eq. (3.4.2), it is clear that we need to know the joint *conditional* distributions for the two tests. If P denotes a population of objects that has the property and NP denotes a population of objects that does not have the property, then we see that the joint distributions of the results in both P and NP are required in order to compute the Bayesian prediction. If the two tests were statistically independent in both P and NP (i.e., statistically conditionally independent) then the two joint probabilities in the right-hand side of Eq. (3.4.2) could be replaced by expressions involving the sensitivities and specificities of the two tests. Namely, if the tests are conditionally independent then the Bayesian prediction can be given by

$$\Pr(P|+_1,+_2) = \frac{1}{1 + \dfrac{\Pr(NP)\Pr(+_1|NP)\Pr(+_2|NP)}{\Pr(P)\Pr(+_1|P)\Pr(+_2|P)}} \quad (3.4.3)$$

where $\Pr(+_i|P)$ ($i = 1, 2$) is the sensitivity of test i and $\Pr(+_i|NP)$ ($i = 1, 2$) is one minus the specificity of test i.

When tests are conditionally independent, the use of Eq. (3.4.3) makes it very simple to compute the Bayesian prediction. One only needs to know how well each individual test works and an initial value for the likelihood that the object has the property [i.e., $\Pr(P)$]. The performance information about the two tests (i.e., the sensitivities and specificities) can come from totally separate sources since knowledge about the tests' joint performances would not be needed.

When tests are conditionally dependent, then the joint results of the two tests on the same set of objects are needed in order to estimate $\Pr(+_1, +_2|P)$ and $\Pr(+_1, +_2|NP)$ in Eq. (3.4.2). In this case two tests with given individual performances (i.e., sensitivities and specificities) can give a wide range of values for the prediction based on two positive results, depending on the type and magnitude of dependence between the two tests. For two tests with estimates for the sensitivities and specificities given by α_i^+ and α_i^- ($i = 1, 2$), we find that the maximum value that $\Pr(+_1, +_2|P)$ can attain is the minimum of the two sensitivities, the minimum that $\Pr(+_1, +_2|P)$ can attain is either the sum of the two sensitivities minus 1 or else zero if the preceding expression is negative in value, the maximum that $\Pr(+_1, +_2|NP)$ can attain is the minimum of $(1 - \alpha_1^-)$ and $(1 - \alpha_2^-)$, and the minimum that $\Pr(+_1, +_2|NP)$ can attain is either $(1 - \alpha_1^- - \alpha_2^-)$ or zero if the value of the preceding expression is negative (Pet-Edwards, 1986).

Two tests are said to be positively dependent if, given the result of one of the tests, the second test is more likely to give the same result than if the tests were independent. A mathematical expression for positive dependence between T_1 and T_2 can be given as follows:

$$\Pr(+_1, +_2) > \Pr(+_1)\Pr(+_2) \quad \text{or} \quad \Pr(-_1, -_2) > \Pr(-_1)\Pr(-_2)$$

Similarly, if $\Pr(+_1, +_2) < \Pr(+_1)\Pr(+_2)$ or $\Pr(-_1, -_2) < \Pr(-_1)\Pr(-_2)$, then the two tests are said to be negatively dependent.

Pairs of tests that have either of the joint *conditional* probabilities given in Eq. (3.4.2) close to their upper bounds are said to be positively conditionally associated or dependent [i.e., if one test gives a particular result, the second test is more likely to give the same result in the particular population (P or NP) than if the tests had been independent]. Tests that have these probabilities closer to their lower bounds are said to be negatively conditionally associated or dependent (e.g., if one test gives a positive result, the second test is less likely to give a positive result than if the tests had been independent). Tests that have $\Pr(+_1, +_2|P) = \Pr(+_1|P)\Pr(+_2|P)$ or

Preliminary Analysis

$\Pr(-_1, -_2 | NP) = \Pr(-_1 | NP)\Pr(-_2 | NP)$ are said to be conditionally independent in P or NP, respectively.

In Fig. 3.4.1, we illustrate the potential range of values that $\Pr(P|+_1, +_2)$ given by Eq. (3.4.2) can assume for different types (positive or negative) and amounts of conditional dependence when both tests give positive results. In this illustration we have used two tests that both have sensitivities and specificities of 0.7. The figure shows the posterior Bayes' probability over the full range of prior probabilities [i.e., Pr(P)] for various types and amounts of conditional dependence. Note that if the two tests are strongly positively dependent in both P and NP (Case C in Fig. 3.4.1), then the prediction based on two positive results is approximately the same as that for a single positive test result and provides a weaker indication of the property than independent tests (Case D in Fig. 3.4.1). However, weak positive dependence in NP and strong positive dependence in P would produce a stronger prediction of P than conditionally independent tests; and strong positive dependence in NP and weak positive dependence in P would produce a weaker prediction of the property than a single test and could even result in a reversal of the prediction (i.e., two positive results could actually indicate a stronger probability that P is absent rather than present). Tests

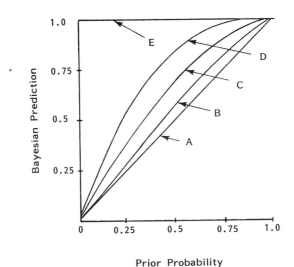

FIGURE 3.4.1. An illustration of the impact of various types and amounts of conditional dependence on Bayesian predictions based on two positive test results. The two tests in the figure both have sensitivities and specificities of 0.7. A, prediction based on no tests (prediction = prior); B, positive dependence in NP and negative in P; C, positive dependence in both P and NP; D, the two tests are conditionally independent; E, maximum negative dependence in NP.

that have negative dependence in NP and positive dependence in P will always provide a stronger indication of the presence of P when two positive results are obtained than conditionally independent tests. Tests that have positive dependence in NP and negative dependence in P will always produce a weaker prediction of P than conditionally independent tests (Case B in Fig. 3.4.1). And finally, if the two tests have strong negative dependence in both P and NP, the prediction of P will be stronger than for independent tests (Case E in Fig. 3.4.1); however, weak negative dependence in NP can make the prediction weaker than for independent tests and can sometimes reverse the prediction.

This example clearly indicates that the "type of the dependence" in each population (i.e., positive or negative conditional dependence) and the amount of conditional dependence can greatly influence the prediction for two positive results given by tests with fixed sensitivities and specificities. Thus far we have only examined the predictions based on two positive results. If we consider all results (i.e., $++$, $+-$, $-+$, and $--$) for a range of different sensitivities and specificities, then the following observations can be made (from Pet-Edwards, 1986): (1) conditionally independent tests provide strong predictions for matching test results, but are generally inconclusive when mixed results are obtained; (2) tests that are positively dependent in one population and negatively dependent in the other provide generally strong predictions regardless of the results of the two tests; (3) negative dependence in both populations provides strong predictions (better than independent tests) when the results are matching, but are generally inconclusive for mixed results; (4) strong positive dependence in both populations generally gives weaker predictions regardless of the test results; and (5) strong positive dependence in one population and weak positive dependence in the other provides relatively strong predictions.

These results clearly indicate that making a decision based on the numerical results of Eq. (3.4.3) (i.e., Bayes' formula for two independent tests) could be greatly improved if information about test dependency is available. The benefit would be even greater if more than two tests were considered. To illustrate the impact of dependence on Bayesian predictions when more than two tests are utilized, examples based on a sequence of one to ten tests that have each given positive results are used. In the case where we are considering n positive test results, Bayes' formula can be given as follows:

$$\Pr(P|+_1, +_2, \ldots, +_n) = \frac{1}{1 + \dfrac{\Pr(NP)\Pr(+_1, +_2, \ldots, +_n | NP)}{\Pr(P)\Pr(+_1, +_2, \ldots, +_n | P)}} \quad (3.4.4)$$

For n tests with estimates for the sensitivities and specificities given by α_i^+

Preliminary Analysis

and α_i^- ($i = 1, 2, \ldots, n$), respectively, we find that the maximum value that $\Pr(+_1, +_2, \ldots, +_n | P)$ in Eq. (3.4.4) can attain is the minimum of the sensitivities; the minimum that $\Pr(+_1, +_2, \ldots, +_n | P)$ can attain is either the sum of the sensitivities minus the quantity $(n - 1)$ or else zero if the preceding expression is negative in value; the maximum that $\Pr(+_1, +_2, \ldots, +_n | NP)$ can attain is the minimum of $(1 - \alpha_1^-)$, $(1 - \alpha_2^-), \ldots, (1 - \alpha_n^-)$; and the minimum that $\Pr(+_1, +_2, \ldots, +_n | NP)$ can attain is either $(1 - \alpha_1^- - \alpha_2^- - \cdots - \alpha_n^-)$ or zero if the value of the preceding expression is negative.

If the prior probability—$\Pr(P)$—is fixed in Eq. (3.4.4), then we can use the preceding bounds to compute the range of predictions [i.e., the range of values for $\Pr(P | +_1, +_2, \ldots, +_n)$] that one could obtain as a function of the number of tests in the battery. Figures 3.4.2–3.4.5 provide four illustrations of this. In each example we have used a prior probability of 0.5; and we show the prediction that would be obtained if the given tests were conditionally independent, as well as the range of predictions for various amounts and types of dependence. In Fig. 3.4.2 we illustrate the bounds on the Bayesian predictions for up to ten tests, each having sensitivities and specificities of 0.7. Figure 3.4.3 illustrates tests having sensitivities of 0.7 and specificities of 0.9; Fig. 3.4.4 illustrates tests having sensitivities of 0.9 and specificities of 0.7; and Fig. 3.4.5 illustrates tests having sensitivities and specificities of 0.9. These figures clearly indicate that obtaining information about the dependencies among the tests becomes more critical as the number of tests in the battery increases (note how wide the bounds on

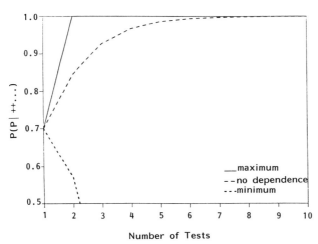

FIGURE 3.4.2. Range of predictions for ten tests with sensitivities of 0.7 and specificities of 0.7. A prior probability of 0.5 is assumed. $[P(P|++\cdots)$ for $P(+|P) = 0.7$ and $P(-|NP) = 0.7$.]

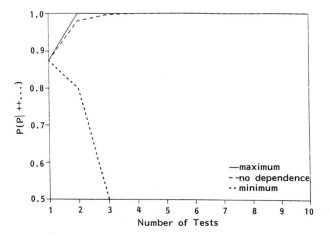

FIGURE 3.4.3. Range of predictions for 10 tests with sensitivities of 0.7 and specificities of 0.9. A prior probability of 0.5 is assumed. [P(P|++···) for P(+|P) = 0.7 and P(−|NP) = 0.9.]

$\Pr(P|+_1, +_2, \ldots, +_n)$ become as the number of tests increases), and that the poorer the performances of the individual tests the wider the bounds (compare the bounds in Figs. 3.4.2–3.4.5). To make Bayesian predictions, it is therefore recommended that, in addition to the performances of the individual tests, information about whether the tests are dependent, including their types (i.e., positive or negative conditional dependence) and magnitudes should also be obtained.

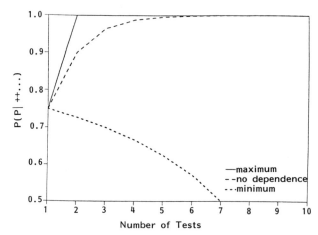

FIGURE 3.4.4. Range of predictions for 10 tests with sensitivities of 0.9 and specificities of 0.7. A prior probability of 0.5 is assumed. [P(P|++···) for P(+|P) = 0.9 and P(−|NP) = 0.7.]

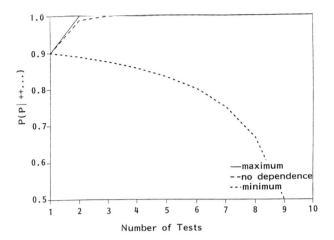

FIGURE 3.4.5. Range of predictions for 10 tests with sensitivities of 0.9 and specificities of 0.9. A prior probability of 0.5 is assumed. [$P(P|++\cdots)$ for $P(+|P) = 0.9$ and $P(-|NP) = 0.9$.]

3.4.2. Testing for Dependencies between Pairs of Tests

We have seen that dependencies among tests can greatly influence the interpretation of the test results. Our first step would be to determine whether any dependencies exist between pairs of tests in the data base. If dependencies are not found, then the Bayesian predictions can be computed using simplified equations involving the sensitivities and specificities of the individual tests, such as Eq. (3.4.3). On the other hand, if dependencies are found between pairs of tests, it will be necessary to estimate the type and amount of conditional dependence so that the Bayesian prediction can be adjusted accordingly.

Two tests, T_1 and T_2, are said to be statistically independent if the following expression holds for all results r_1 and r_2 of T_1 and T_2, respectively:

$$\Pr(r_1, r_2) = \Pr(r_1)\Pr(r_2) \qquad (3.4.5)$$

If the equality in Eq. (3.4.5) does not hold, then the tests are said to be statistically dependent.

In order to check whether a pair of tests is statistically dependent using information contained in a data base, we must generally contend with a limited number of test results. It is therefore not enough to simply compute estimates for the probabilities in Eq. (3.4.5) and then to check whether or not the equality holds. We must also account for the quality and quantity of the data. For example, suppose from the data base that we compute the following estimates: $\Pr(+_1, -_2) = 0.35$ and $\Pr(+_1)\Pr(-_2) = 0.30$. Since the two estimates are not equal, is this enough to "conclude" that the two tests are statistically dependent? The answer to this question depends on how many joint test results there were in the data base to compute the estimates. If the number of test results were "small enough," then the difference between 0.35 and 0.30 would not be significant. On the other hand, if the number of results were "large enough" then the difference would be significant. Therefore, in order to check whether dependencies exist between pairs of tests, we need a statistic that will measure the significance of deviations from Eq. (3.4.5).

There are a number of statistics that can be used to check for dependencies among tests. The oldest and perhaps most widely used class of statistics (especially in the social sciences) is that which is based on the chi-square statistic. This class of dependence statistics includes, in addition to the standard chi-square statistic itself, the phi-square (which is chi-square divided by the total sample size), Pearson's coefficient of contingency (Pearson, 1904), Tschuprow's contingency coefficient (Tschuprow, 1919), and Cramer's contingency coefficient (Cramer, 1946). These measures were designed for inferential uses and focus on the "nonrelation" statistical independence. In other words, they provide an indication of the likelihood that a given pair of tests is statistically independent based on the joint test results. They also form a measure of the deviation (but not direction) from statistical independence. If independence fails to hold, the sizes of these statistics tend to reflect the squared deviations of the population probabilities from independence.

In the CPBS method, we use the chi-square statistic (in addition to the results from cluster analysis) in order to obtain indications of whether dependencies exist between pairs of tests. We compute this statistic for each pair of tests in the data base. In particular, we construct contingency tables of the form given in Table 3.4.1 for each pair of tests, T_i and T_j.

We then divide each cell entry in Table 3.4.1 by n (i.e., the number of joint test results), and the table given in Table 3.4.2 is produced. Using the information in Table 3.4.2, we can compute a chi-square statistic with a continuity correction (see Fleiss, 1981) to test for dependence between test i and test j, as follows:

$$\chi^2 = n \sum_{i=1}^{2} \sum_{i=1}^{2} \frac{(|P_{ij} - P_{i.}P_{.j}| - 1/2n)^2}{P_{i.}P_{.j}} \tag{3.4.6}$$

Preliminary Analysis

TABLE 3.4.1. Contingency Table between Tests i and j[a]

		Test i		
		+	−	Row sum
Test j	+	n_{11}	n_{12}	$n_{1.}$
	−	n_{21}	n_{22}	$n_{2.}$
Column sum		$n_{.1}$	$n_{.2}$	n

[a] n_{11} is the number of objects where test i shows + and test j shows +; n_{21} is the number of objects where test i shows + and test j shows −; n_{12} is the number of objects where test i shows − and test j shows +; n_{22} is the number of objects where test i shows − and test j shows −; and n is the total number of objects tested by both test i and test j.

This statistic can be used to determine whether the given pair of tests is statistically independent (as long as the entries in the table are "sufficiently large"). If the chi-square value computed in Eq. (3.4.6) is found to exceed a given reference value found in a table of chi-square values with one degree of freedom, then we would conclude that the pair of tests shows statistical *independence* at the given percentage level. (For convenience, a portion of the chi-square table for one degree of freedom is given in Table 3.4.3.)

In order for the reader to better understand how the chi-square statistic can be interpreted, we provide the following example: Suppose for a given pair of tests we compute χ^2 to be 3.90 using Eq. (3.4.6). By reference to Table 3.4.3 we can see that there is a 5% chance that the chi-square value would exceed 3.84 if the two tests were independent (i.e., for a value of χ^2 of 3.84 or greater, statistical *independence* is significant at the 0.05 level). Thus we could conclude with a fairly high degree of confidence that the two tests are not independent (i.e., the tests are very likely to be dependent).

TABLE 3.4.2. Contingency Table Used in Computing the Chi-Square Statistic[a]

		Test i		
		+	−	Row sum
Test j	+	P_{11}	P_{12}	$P_{1.}$
	−	P_{21}	P_{22}	$P_{2.}$
Column sum		$P_{.1}$	$P_{.2}$	1.0

[a] Computed from Table 3.4.1 by dividing each entry by n.

TABLE 3.4.3. Percentage Points of
Chi-Square Distribution for One
Degree of Freedom

Percentage	Chi-square value
0.005	0.0000393
0.010	0.000157
0.025	0.000982
0.050	0.00393
0.100	0.0158
0.250	0.102
0.500	0.455
0.750	1.32
0.900	2.71
0.950	3.84
0.975	5.02
0.990	6.63
0.995	7.88

We indicated above that the results of the chi-square test are only valid (i.e., the results are significant) if the entries in the table are "sufficiently large." A criterion that one can employ to determine whether the computed chi-square statistic is significant is to examine the value of the *minimum expected frequency* of the cells. Statistical independence implies that the P_{ij}'s in Table 3.4.2 are (approximately) equal to the products $(P_{i.})(P_{.j})$ ($i = 1, 2$ and $j = 1, 2$). If a pair of tests is statistically independent, then the "expected frequency" in cell (ij) will be given by the product $n(P_{i.})(P_{.j})$, where n represents the number of joint test results in the data base between the given pair of tests. In order for the chi-square statistic to be significant, these expected frequencies must not be too small. This vague term—"too small"—has generally been interpreted as meaning that all expected frequencies in the cells must be greater than 5 (alternatively, the minimum of the expected frequencies must be greater than 5). Several authors have indicated that the chi-square statistic is still valid for expected frequencies as low as 0.5 (Everitt, 1977).

As an illustration of the computations, consider the following hypothetical example: Suppose we are considering two tests in our data base, say test A and test B, and we want to evaluate whether the tests are dependent or independent. From an examination of their joint responses, we observe that both tests gave positive results on 60 objects, negative results on 25 objects, test A was positive and test B was negative on 5 objects, and test A was negative and test B was positive on 5 objects. These results can be summarized in the following table:

Preliminary Analysis

	Test A +	Test A −	Row sum
Test B +	60	5	70
−	5	25	30
Column sum	70	30	100

If we divide each entry of the preceding table by the number of objects tested, namely 100, then we get the following table (the reader should also refer to Table 3.4.2):

	Test A +	Test A −	Row sum
Test B +	0.60	0.05	0.7
−	0.05	0.25	0.3
Column sum	0.70	0.30	1.0

The chi-square statistic given in Eq. (3.4.6) would thus be computed as the following sum:

$$100\left[\frac{(|0.6-(0.7)(0.7)|-1/200)^2}{(0.7)(0.7)} + \frac{(|0.05-(0.7)(0.3)|-1/200)^2}{(0.7)(0.3)}\right.$$

$$\left. + \frac{(|0.05-(0.3)(0.7)|-1/200)^2}{(0.3)(0.7)} + \frac{(|0.25-(0.3)(0.3)|-1/200)^2}{(0.3)(0.3)}\right]$$

$$\chi^2 = 100(0.0225 + 0.114 + 0.114 + 0.267) = 51.75$$

The minimum expected frequency would be given by

$$\min\{100(0.7)(0.7), 100(0.7)(0.3), 100(0.3)(0.7), 100(0.3)(0.3)\}$$

$$= \min\{49, 21, 21, 9\} = 9$$

The value for the minimum expected frequency (i.e., 9) indicates that the results of the chi-square test can be considered as significant; and the value of 51.75 for χ^2 indicates that there is a less than 0.5% chance that the tests are independent (refer to Table 3.4.3). Thus, we can feel fairly confident that tests A and B are dependent.

The following example based on the cancer hazard identification problem adapted from Pet-Edwards *et al.* (1985a) is used to further illustrate the computation of the chi-square statistic.

EXAMPLE 3.4.1

In the Gene-Tox data base (see Palajda and Rosenkranz, 1985), there are 11 tests for which both the sensitivities and specificities of the tests can be estimated. These tests include Sty, PrM, EcW, DRL, UDS, SHE, BHK, HMA, Cvt, SCE, and VET. (The reader should refer to the Appendix for an explanation of the test abbreviations.) These 11 tests can be further analyzed in pairs to determine whether any dependencies exist between them. Table 3.4.4 gives the results of the chi-square test given in Eq. (3.4.6) along with the minimum expected frequency for each pair of tests. We can see that the minimum expected frequencies for many pairs of tests are small (i.e., less than 0.5). For these pairs, we should not use the chi-square statistic to draw conclusions about dependencies. From the "significant" results we see that there is a fairly strong indication that the pairs of tests (Sty, EcW), (Sty, DRL), (Sty, UDS), (EcW, DRL), and (DRL, UDS) are dependent (the probability that they are independent is less than 0.05), and a slightly weaker indication that the pairs (Sty, HMA) and (EcW, UDS) are dependent (the probability that they are independent is less than 0.10).

In the cases where the minimum expected frequencies are small and consequently the results of the chi-square test should not be considered significant, we can also examine the results of cluster analysis using simple matching coefficients to see whether any pairs of tests show *positive* associations or dependencies (see Section 3.3). Note that the chi-square test can indicate the presence of dependencies, both positive and negative, but cannot distinguish between the two types of dependence, whereas, cluster analysis can provide an indication of the presence of positive dependencies. Two tests are found within the same cluster if they have given similar responses on the same set of objects. If this level of similarity is 100%, then it is obvious that the tests are strongly positively dependent. If the level of similarity is very high but less than 100%, then the tests show moderate positive dependence. (The reader should refer to Section 3.3 for a more detailed discussion of this.)

We can use the results of the spanning tree method together with the single-link method (see Sections 2.2 and 3.3) to establish which pairs of tests are most similar. The following example, adapted from Pet-Edwards *et al.* (1985a, b) is used to illustrate this use of cluster analysis.

TABLE 3.4.4. Results of Chi-Square Test on the Gene-Tox Data Base[a]

Assays		Chi square	Minimum expected frequency	Significance
Sty	PrM	4.120	0.444[b]	—
Sty	EcW	19.157	6.533	0.025[c]
Sty	DRL	19.269	4.267	0.05[c]
Sty	UDS	24.083	10.000	0.005[c]
Sty	SHE	1.212	2.240	0.25
Sty	BHK	2.760	1.806	0.25
Sty	HMA	3.061	3.291	0.10[d]
Sty	Cvt	0.024	0.429[b]	—
Sty	SCE	1.998	0.581	0.50
Sty	VET	0.155	2.066	0.25
PrM	EcW	2.939	0.571	0.50
PrM	DRL	308.580	0.000[b]	—
PrM	UDS	0.696	0.200[b]	—
PrM	SHE	124253.121	0.000[b]	—
PrM	BHK	0.439	0.250[b]	—
PrM	HMA	0.376	0.250[b]	—
PrM	Cvt	498.502	0.000[b]	—
PrM	SCE	285.828	0.000[b]	—
PrM	VET	82504.077	0.000[b]	—
EcW	DRL	6.465	4.263	0.05[c]
EcW	UDS	9.931	3.150	0.10[d]
EcW	SHE	0.496	0.222[b]	—
EcW	BHK	285.828	0.000[b]	—
EcW	HMA	3.581	1.067	0.50
EcW	Cvt	49106.040	0.000[b]	—
EcW	SCE	0.788	0.182[b]	—
EcW	VET	2.398	0.600	0.50
DRL	UDS	14.063	4.000	0.025[c]
DRL	SHE	1.234	0.133[b]	—
DRL	BHK	262.030	0.000[b]	—
DRL	HMA	0.165	0.300[b]	—
DRL	Cvt	0.863	0.167[b]	—
DRL	SCE	3.219	0.071[b]	—
DRL	VET	1.158	0.136[b]	—
UDS	SHE	0.060	1.300	0.50
UDS	BHK	0.207	0.294[b]	—
UDS	HMA	2.720	1.842	0.25
UDS	Cvt	148.250	0.000[b]	—
UDS	SCE	246.420	0.000[b]	—
UDS	VET	0.055	1.714	0.25

(*continued*)

TABLE 3.4.4. (*Continued*)

Assays		Chi square	Minimum expected frequency	Significance
SHE	BHK	5.867	0.316[b]	—
SHE	HMA	0.000	0.500	0.50
SHE	Cvt	157.114	0.000[b]	—
SHE	SCE	13753.275	0.000[b]	—
SHE	VET	2.991	0.462[b]	—
BHK	HMA	2.553	0.522	0.50
BHK	Cvt	2.056	0.100[b]	—
BHK	SCE	3.206	0.067[b]	—
BHK	VET	1.201	0.143[b]	—
HMA	Cvt	0.299	0.267[b]	—
HMA	SCE	2.723	0.074[b]	—
HMA	VET	0.014	0.462[b]	—
Cvt	SCE	134.288	0.000[b]	—
Cvt	VET	157.114	0.000[b]	—
SCE	VET	1.854	0.100[b]	—

[a] Taken from Pet-Edwards *et al.* (1985a).
[b] Expected frequency is too small for chi square to be conclusive.
[c] Strong indication that the pair of tests is not independent.
[d] Moderate indication that the pair of tests is not independent.

EXAMPLE 3.4.2

The results of the spanning tree and single-link method on the Gene-Tox data base using simple matching coefficients were presented earlier in Figs. 3.3.7 and 3.3.5, respectively, in Section 3.3. Focusing on the 11 tests listed in Example 3.4.1, the following list of (potentially) positively dependent test pairs (having similarity levels of 90% or greater) can be derived:

Percent similarity	Test pair
100	PrM, EcW
96	SHE, BHK
92	HMA, SCE
92	Cvt, SCE
90	Sty, EcW

The results of the chi-square test and cluster analysis can indicate which of the tests are likely to be dependent. This information is useful in battery selection, where one might want to try to select tests that are independent, and when making Bayesian predictions. When one wants to make predictions based on dependent tests, then measures of the amount of *conditional dependence* between the tests are needed. The following describes the conditional dependence measure that we have developed for use in the CPBS.

3.4.3. Measuring the Conditional Dependence between Tests

As illustrated earlier, both the direction and amount of conditional dependence can influence the value of the Bayesian predictions. Consequently, we have found that the following characteristics for measures of conditional dependence would be useful if we want to use the measures in Bayes' formula: (1) the measures should indicate the direction (positive or negative) of conditional dependence; (2) the measures should indicate the amounts of departure from independence; and (3) knowledge of the measures together with the sensitivities and specificities of the tests should make Bayesian predictions possible and convenient to compute. It was found that none of the traditional dependence measures had all of the desirable characteristics for use in the CPBS, and a new set of measures of conditional dependence was developed for use in the CPBS (see Pet-Edwards, 1986). Its development began with an examination of the definition of conditional independence between two tests:

Definition. Two tests are said to be mutually conditionally independent if and only if the following holds for all joint outcomes—r_1 and r_2—of the tests:

$$\Pr(r_1, r_2 | P) = \Pr(r_1 | P) \Pr(r_2 | P)$$

$$\Pr(r_1, r_2 | NP) = \Pr(r_1 | NP) \Pr(r_2 | NP)$$

Tests T_1 and T_2 are said to be conditionally dependent given the presence or absence of property P if either of the preceding equations does not hold.

The preceding equations were examined to see how they could be best adjusted to account for the dependence between a pair of tests, and it was found that a multiplicative adjustment was convenient. This resulted in the following set of measures for conditional dependence.

Conditional Dependence Measures

$$K_P(r_1, r_2) = \frac{\Pr(r_1, r_2 | P)}{\Pr(r_1 | P)\Pr(r_2 | P)} \quad (3.4.7)$$

$$K_{NP}(r_1, r_2) = \frac{\Pr(r_1, r_2 | NP)}{\Pr(r_1 | NP)\Pr(r_2 | NP)} \quad (3.4.8)$$

where $K_P(r_1, r_2)$ is the amount of conditional dependence in P and $K_{NP}(r_1, r_2)$ is the amount of conditional dependence in NP.

Thus, in the case where a pair of tests was not independent, the independence relations would be adjusted as follows:

$$\Pr(r_1, r_2 | P) = K_P(r_1, r_2)\Pr(r_1 | P)\Pr(r_2 | P)$$

and

$$\Pr(r_1, r_2 | NP) = K_{NP}(r_1, r_2)\Pr(r_1 | NP)\Pr(r_2 | NP)$$

Note that a value of one for $K_P(r_1, r_2)$ or $K_{NP}(r_1, r_2)$ would indicate conditional independence, a value greater than one (for matching results) would indicate positive conditional dependence, and a value of less than one (for matching results) would indicate negative conditional dependence. Also, the more the dependence measures would deviate from the value one, the greater the conditional dependence would be.

If we substitute the expressions for the joint probabilities given in Eqs. (3.4.7) and (3.4.8) into Eq. (3.4.2) we can see how the dependence measures can be conveniently incorporated into Bayes' formula when tests are dependent:

$$\Pr(P | +_1, +_2) = \frac{1}{1 + \dfrac{K_{NP}(+_1, +_2)\Pr(+_1 | NP)\Pr(+_2 | P)\Pr(NP)}{K_P(+_1, +_2)\Pr(+_1 | P)\Pr(+_2 | P)\Pr(P)}}$$

Thus we can see that this set of dependence measurements satisfies the three desirable characteristics described earlier. The remainder of this section is used to discuss and illustrate how the conditional dependence measures can be estimated and how the significance of conditional dependence can be tested.

Notice that if we are interested in the conditional dependence between two positive results in P, then the denominator in Eq. (3.4.7) is given by the product of the sensitivities of the two tests; and similarly if we are interested in the conditional dependence between two negative results in NP, then the denominator in Eq. (3.4.8) is given by the product of the specificities of the two tests. The estimates for the sensitivities would be

Preliminary Analysis

given by the proportion of the cases that have the property of interest for which each test gave positive results, i.e., α_1^+ and α_2^+; and the estimates for the specificities would be given by the proportion of the cases that do not have the property of interest for which each test gave negative results, i.e., α_1^- and α_2^-. We could thus write the expressions for the conditional dependence measures, hereafter called the "KP measures," as follows:

$$K_P(+_1, +_2) = \frac{\Pr(+_1, +_2 | P)}{(\alpha_1^+)(\alpha_2^+)} \tag{3.4.9}$$

$$K_{NP}(-_1, -_2) = \frac{\Pr(-_1, -_2 | NP)}{(\alpha_1^-)(\alpha_2^-)} \tag{3.4.10}$$

where an estimate for the numerator of Eq. (3.4.9) is given by the proportion of the cases that have the property of interest for which both tests gave positive results; and the numerator of Eq. (3.4.10) is given by the proportion of the cases that do not have the property of interest for which both tests gave negative results.

If we compute the two estimates given in Eqs. (3.4.9) and (3.4.10) from the data, then the KP measures associated with the remaining test results can be computed as functions of these. In particular, the following relationships hold among the pairwise dependence measures (from Pet-Edwards, 1986):

$$K_P(+_1, -_2) = \frac{1 - K_P(+_1, +_2)\alpha_2^+}{1 - \alpha_2^+} \tag{3.4.11}$$

$$K_P(-_1, +_2) = \frac{1 - K_P(+_1, +_2)\alpha_1^+}{1 - \alpha_1^+} \tag{3.4.12}$$

$$K_P(-_1, -_2) = \frac{1 - \alpha_1^+ - \alpha_2^+ + K_P(+_1, +_2)\alpha_1^+\alpha_2^+}{(1 - \alpha_1^+)(1 - \alpha_2^+)} \tag{3.4.13}$$

$$K_{NP}(+_1, -_2) = \frac{1 - K_{NP}(-_1, -_2)\alpha_1^-}{1 - \alpha_1^-} \tag{3.4.14}$$

$$K_{NP}(-_1, +_2) = \frac{1 - K_{NP}(-_1, -_2)\alpha_2^-}{1 - \alpha_2^-} \tag{3.4.15}$$

$$K_{NP}(+_1, +_2) = \frac{1 - \alpha_1^- - \alpha_2^- + K_{NP}(-_1, -_2)\alpha_1^-\alpha_2^-}{(1 - \alpha_1^-)(1 - \alpha_2^-)} \tag{3.4.16}$$

Equations (3.4.9) and (3.4.10) are point estimates for the true conditional dependence between two tests. The sample variances of these estimates can be given as follows: Suppose we assume that the sensitivities and specificities in Eqs. (3.4.9) and (3.4.10) are fixed and known. Then the joint probabilities in the two equations have an approximate binomial distribution. The following estimate for the variance of the KP measures can thus be derived (from Pet-Edwards, 1986):

$$\text{var}[K_P(+_1, +_2)] = \frac{\Pr(+_1, +_2 | P)(1 - \Pr(+_1, +_2 | P))}{n_P(\alpha_1^+)(\alpha_2^+)} \quad (3.4.17)$$

$$\text{var}[K_{NP}(-_1, -_2)] = \frac{\Pr(-_1, -_2 | NP)(1 - \Pr(-_1, -_2 | NP))}{n_{NP}(\alpha_1^-)(\alpha_2^-)} \quad (3.4.18)$$

where n_P is the sample size in P and n_{NP} is the sample size in NP.

Given the two point estimates in Eqs. (3.4.9) and (3.4.10) and their associated variances in Eqs. (3.4.17) and (3.4.18), we can test whether each of the KP dependence estimates is significantly different from the value one (i.e., significantly different from independence). In particular, assuming that n_P and n_{NP} are reasonably large, we compute the following statistics and compare the results to tables of the standard normal distribution:

$$Z_P = \frac{|K_P(+_1, +_2) - 1|}{[\text{var}(K_P(+_1, +_2))]^{1/2}} \quad (3.4.19)$$

$$Z_{NP} = \frac{|K_{NP}(-_1, -_2) - 1|}{[\text{var}(K_{NP}(-_1, -_2))]^{1/2}} \quad (3.4.20)$$

A value of Z_P or Z_{NP} of greater than $Z_{\beta/2}$ found in a table of Standard Normals, would mean that we are $100(1 - \beta)\%$ confident that the KP measure in P or NP, respectively, is significantly different from independence. For example, a value of Z_P or Z_{NP} that is greater than 1.645 would represent that the two tests are conditionally dependent with a 90% confidence level, a value greater than 1.96 would represent a 95% confidence level, and a value greater than 2.576 would represent a 99% confidence level.

The computation of conditional dependence measures is illustrated in the following example, which uses a data base recently reported by the National Toxicology Program (NTP) (Tennant et al., 1987).

EXAMPLE 3.4.3

The data base consists of 73 chemicals evaluated by a standard protocol for carcinogenicity and for genotoxicity in four short-term tests: *Salmonella* mutagenicity (Sty), chromosomal aberrations *in vitro* (Cvt), sister chromatid exchanges (SCE), and gene mutations in mouse lymphoma L5178Y cells

TABLE 3.4.5. List of Chemicals and Test Results in the NTP Data Base

CAS number	Chemical name	Ca	Sty	Cvt	SCE	Mly
00057-06-7	Allyl isothiocyanate	+	−	+	+	+
02835-39-4	Allyl isovalerate	+	−	+	+	+
02432-99-7	11-Aminoundecanoic acid	+	−	−	+	−
00050-81-7	L-Ascorbic acid	−	−	−	+	−
00071-43-2	Benzene	+	−	−	+	−
00119-53-9	Benzoin	−	−	−	−	+
00140-11-4	Benzyl acetate	+	−	−	−	+
02185-92-4	2-Biphenylamine hydrochloride	+	+	+	−	+
00108-60-1	Bis(2-chloro-1-methylethyl)ether	+	+	+	+	+
00080-05-7	Bisphenol A	−	−	−	−	−
00085-68-7	Butyl benzyl phthalate	+	−	−	−	−
00105-60-2	Caprolactam	−	−	−	−	−
00108-90-7	Chlorobenzene	−	−	−	+	+
00124-48-1	Chlorodibromomethane	+	−	−	+	+
00107-07-3	2-Chloroethanol	−	+	+	+	+
00563-47-3	3-Chloro-2-methyl propene	+	−	+	+	+
01936-15-8	C.I. Acid Orange 10	−	−	+	−	−
03567-69-9	C.I. Acid Red 14	−	−	−	−	−
00518-47-8	C.I. Acid Yellow 73	−	−	−	+	+
02832-40-8	C.I. Disperse yellow 3	+	+	−	+	+
00842-07-9	1-Phenylazo-2-naphthol	+	+	−	+	+
00087-29-6	Cinnamyl anthranilate	+	−	−	−	+
21739-91-3	Cytembena	+	+	+	+	+
05160-02-1	D and C Red No. 9	+	+	−	−	−
00131-17-9	Diallyl phthalate	−	−	+	+	+
00096-12-8	1,2-Dibromo-3-chloropropane	+	+	+	+	+
00106-93-4	Ethylene dibromide	+	+	+	+	+
00095-50-1	1,2-Dichlorobenzene	−	−	−	+	+
00609-20-1	2,6-Dichloro-*p*-phenylenediamine	+	+	+	+	+
00078-87-5	Propylene chloride	+	+	+	+	+
00542-75-6	1,3-Dichloropropene	+	+	−	+	+
00103-23-1	Di(2-ethylhexyl)adipate	+	−	+	−	−
00117-81-7	Di(2-ethylhexyl)phthalate	+	−	−	+	−
00101-90-6	Diglycidylresorcinol ether	+	+	+	+	+
00868-85-9	Dimethyl hydrogenphosphite	+	+	+	+	+
00597-25-1	Dimethyl morpholino-phosphamidate	+	−	+	+	+
00120-61-6	Dimethyl 1,4-benzene-dicarboxylate	−	−	−	−	−
09002-92-0	Ethoxylated dodecyl alcohol	−	−	−	−	−
00140-88-5	Ethyl acrylate (inhibited)	+	−	+	+	+
00097-53-0	Eugenol	−	−	+	+	+
02783-94-0	Sunset yellow FCF	−	−	−	−	+
00105-87-3	Geranyl acetate	−	−	−	+	+

(*continued*)

TABLE 3.4.5. (*Continued*)

CAS number	Chemical name	Ca	Sty	Cvt	SCE	Mly
68916-39-2	Witch hazel	−	−	−	−	−
02784-94-3	HC Blue No. 1	+	+	+	+	+
33229-34-4	HC Blue 2	−	+	−	+	+
00148-24-3	8-Hydroxyquinoline	−	+	+	+	+
00078-59-1	Isophorone	+	−	−	+	+
01634-78-2	Malaoxone	−	−	−	+	+
00069-65-8	Mannitol	−	−	−	−	−
00108-78-1	Melamine	+	−	−	−	−
15356-70-4	DL-Menthol	−	−	−	−	−
13552-44-8	4,4′-Methylenedianiline diHCl	+	+	+	+	+
00150-68-5	Monuron	+	−	+	+	−
00101-80-4	Bis(4-Aminophenyl)ether	+	+	+	+	+
00076-01-7	Pentachloroethane	+	−	−	+	+
00108-95-2	Phenol	−	−	+	+	+
67774-32-7	Polybrominated biphenyls	+	−	−	−	−
00121-79-9	Propyl gallate	−	−	+	+	+
00075-56-9	Propylene oxide	+	+	+	+	+
00050-55-5	Reserpine	+	−	−	−	−
07446-34-6	Selenium sulfide	+	+	+	+	+
00126-92-1	Sodium(2-ethylhexyl)alcohol sulfate	−	−	−	−	−
07772-99-8	Stannous chloride	−	−	+	+	−
00127-69-5	Sulfisoxazole	−	−	−	+	+
01746-01-6	2,3,7,8-TCDD	+	−	−	−	−
00630-20-6	1,1,1,2-Tetrachloroethane	+	−	−	+	+
13463-67-7	Titanium dioxide	−	−	−	−	−
26471-62-5	2,4- and 2,6-Toluene diisocyanate	+	+	+	+	+
15481-70-6	Toluene-2,6-diamine diHCl	−	+	+	+	+
00079-01-6	1,1,2-Trichloroethylene	+	−	−	+	+
00078-42-2	Tris(2-ethylhexyl)phosphate	+	−	−	−	−
17924-92-4	Zearalenone	+	−	+	+	−
00137-30-4	Ziram	+	+	+	−	+

(Mly). Results are listed in Table 3.4.5, treating equivocal results as negative, as was done in the original publication of Tennant *et al.* (1987). Sensitivities and specificities of the four assays are listed in Table 3.4.6. The conditional dependence measures were computed using Eqs. (3.4.9) and (3.4.10), the variances were computed using Eqs. (3.4.17) and (3.4.18), and the significances of the measures were computed using Eqs. (3.4.19) and (3.4.20). These results are summarized in Table 3.4.7. All of the dependence measures are greater than one, indicating that each pair of assays is positively dependent, although only about half of the pairs of tests show deviations from independence at a significance level of 90% or more. When $K_P(++)$ is greater than one, $K_P(--)$ must also be greater than one, and similarly,

Preliminary Analysis

TABLE 3.4.6. Sensitivities and Specificities
Computed from the Data in Table 3.4.5

Test	Sensitivity (α^+)	Specificity (α^-)
Sty	0.455	0.862
Cvt	0.545	0.690
SCE	0.727	0.448
Mly	0.705	0.448

when $K_{NP}(--)$ is greater than one, $K_{NP}(++)$ must also be greater than one, as can be easily shown from Eqs. (3.4.13) and (3.4.14). Thus, the positive dependence means that there is a greater than expected agreement between each pair of assays (than if they were independent), and that this agreement can be correct (i.e., positive results for a carcinogen or negative results for a noncarcinogen) or incorrect (i.e., negative results for a carcinogen or positive results for a noncarcinogen). This observation supports the concept that there is a common property of "genotoxicity" which may be measured by all four tests (Ashby and Tennant, 1988), and leads to a number of possibilities for further investigation.

The following is a summary of the steps used in the CPBS to test for dependencies between pairs of tests.

PRELIMINARY ANALYSIS: CPBS PROCEDURE FOR TESTING FOR DEPENDENCIES

Step 1. Apply the chi-square test given in Eq. (3.4.6) on the data base (both objects with known and unknown properties can be used) in order to identify pairs of tests that are potentially statistically dependent.

TABLE 3.4.7. The KP-Dependence Measures Computed from the Data in Table 3.4.5

Tests	Pr(+, +)	Var	K_P	Z_P	Pr(−, −)	Var	K_{NP}	Z_{NP}
Sty, Cvt	0.364	0.0212	1.468	3.21[a]	0.655	0.0131	1.102	0.891
Sty, SCE	0.386	0.0163	1.169	1.32	0.448	0.0221	1.160	1.077
Sty, Mly	0.432	0.0174	1.349	2.64[a]	0.448	0.0221	1.160	1.077
Cvt, SCE	0.477	0.0143	1.203	1.70[c]	0.414	0.0271	1.338	2.053[b]
Cvt, Mly	0.477	0.0148	1.242	1.993[b]	0.379	0.0263	1.227	1.398
SCE, Mly	0.614	0.0105	1.198	1.926[c]	0.379	0.0404	1.887	4.414[a]

[a] 99% significance level.
[b] 95% significance level.
[c] 90% significance level.

Step 2. Use the results of the single-link and spanning tree clustering methods on the data base to help identify positively dependent pairs of tests.

Step 3. Using the portion of the data base containing objects with known properties, compute the K_P and K_{NP} conditional dependence estimates using Eqs. (3.4.7) and (3.4.8), respectively, for each pair of tests. Then check whether the KP measures are significantly different from the value one, using Eqs. (3.4.17)-(3.4.20).

Step 4. Combine the results given in Steps 1-3 to obtain a list of dependent test pairs along with measures of their conditional dependence for use in battery selection and Bayesian prediction.

Note that ideally we would like to obtain both a list of the conditionally dependent tests and the measures of their conditional dependence that can be used in Bayesian prediction and battery selection. The feasibility of doing this is dependent on the amount of raw data that is available for the tests. Steps 1 and 2 provide only a list of dependent test pairs, whereas Step 3 provides the list of conditionally dependent tests as well as the measures. Since we are really after the conditional dependence measures, we can see that Steps 1 and 2 can be omitted if the data base has sufficient numbers of joint results in both populations P and NP. However, if the data base is nonideal, then the measures of dependence in Step 3 may be impossible to compute for some pairs of tests owing to limited data on objects with known properties. In this case, Steps 1 and 2 (on the expanded data base containing objects with both known and unknown properties) can be used to provide indications of dependencies. This information can be used in battery selection to help select batteries that are statistically independent (see Chapter 4). (We noted earlier that simplified expressions for combining test results in Bayes' formula can be derived when the tests are conditionally independent.) Test pairs that are found to be dependent in Steps 1 and 2 can still be used in the Bayesian predictions, but care should be taken in the interpretation of the results, especially when the number of tests in the battery is large and the individual tests have moderate to poor performance.

REFERENCES

Ashby, J., and Tennant, R. W., 1988, "Chemical structure, Salmonella mutagenicity and extent of carcinogenicity as indicators of genotoxic carcinogenesis among 222 chemicals tested in rodents by the U.S. NCI/NTP," *Mutation Res.*, **204**:17-115.
Chankong, V., Haimes, Y. Y., Rosenkranz, H. S., and Pet-Edwards, J., 1985, "The carcinogenicity prediction and battery selection (CPBS) method: A Bayesian approach," *Mutation Res.*, **153**(3):135-166.

Cramer, H., 1946, *Mathematical Methods of Statistics*, Princeton University Press, Princeton, New Jersey.
Everitt, B. S., 1977, *The Analysis of Contingency Tables*, Chapman and Hall, London.
Fleiss, J. L., 1981, *Statistical Methods for Rates and Proportions*, Wiley, New York.
Galen, R. S., and Gambino, S. R., 1975, *Beyond Normality: The Predictive Value and Efficiency of Medical Diagnoses*, Wiley, New York.
Greenes, R. A., Begg, C. B., Cain, K. C., Swets, J. A., Feehrer, C. E., and McNeil, B. J., 1984, "Patient-oriented performance measures of diagnostic tests: 2. Assignment potential and assignment strength," *Med. Decision Making*, 4(1).
IARC Monographs on the Evaluation of the Carcinogenic Risk of Chemicals to Humans, 1982, Suppl. 4, *Chemicals, Industrial Processes and Industries Associated with Cancer in Humans*, International Agency for Research on Cancer, Lyon.
Palajda, M., and Rosenkranz, H. S., 1985, "Assembly and preliminary analysis of a genotoxicity data base for predicting carcinogens," *Mutation Res.*, 153:79–135.
Pearson, K., 1904, "Mathematical contributions to the theory of evolution, XIII On the theory of contingency and its relation to association and normal correlation," *Draper's C. Res. Mem. Biometric*, Sec. 1, Reprinted in *Karl Pearson's Early Papers*, Cambridge University Press, Cambridge, 1948.
Pet-Edwards, J., 1986, "Selection and interpretation of conditionally dependent tests for binary predictions: A Bayesian approach," Ph.D. dissertation, Case Western Reserve University, Cleveland, Ohio.
Pet-Edwards, J., Chankong, V., Rosenkranz, H. S., and Haimes, Y. Y., 1985a, "Application of the CPBS method to the Gene-Tox data base," *Mutation Res.*, 153:187–200.
Pet-Edwards, J., Rosenkranz, H. S., Chankong, V., and Haimes, Y. Y., 1985b, "Cluster analysis in predicting the carcinogenicity of chemicals using short-term assays," *Mutation Res.*, 153:173–192.
Tennant, R. W., Margolin, B. H., Shelby, M. D., Zeiger, E., Haseman, J. K., Spalding, J., Caspary, W., Resnick, M., Stasiewicz, S., Anderson, B., and Minor, R., 1987, "Prediction of chemical carcinogenicity in rodents from *in vitro* genotoxicity assays," *Science* 236:933–941.
Tschuprow, A. A., 1919, "On mathematical expectation of the moments of frequency distributions," *Biometrika*, 12:140–169.
Waters, M. D., Stack, H. F., and Brady, A. L., 1986, "Analysis of the spectra of genetic activity in short-term tests," in *Genetic Toxicology of Environmental Chemicals, Part B: Genetic Effects and Applied Mutagenesis*, C. Ramel, ß. Lambert, and J. Magnusson (eds.), Alan R. Liss, New York, pp. 99–109.
Zeiger, E., 1982, "Knowledge gained from the testing of large numbers of chemicals," in *Environmental Mutagens and Carcinogens, Proceedings of the Third International Conference on Environmental Mutagens*, Tokyo, Mishima, and Kyoto, September 21–27, 1981, T. Sugimura, S. Kondo, and H. Takebe (eds.), Alan R. Liss, New York, pp. 337–344.

Chapter 4

Battery Selection

In Chapter 3 we described several analyses that can be used to summarize the performances of the individual tests and the interdependencies among the tests from a data base containing test results on objects with known properties. With the availability of this summary information, whether it is obtained from the preliminary analyses or from other sources, we are now in a position to try to determine which combination of tests would be best to use for a given decision problem.

For our purposes, the problem of selecting the best battery of tests can be formally defined as follows:

Suppose we are given the following information:

1. A set of n tests;
2. Estimates for the sensitivities and specificities of the n tests;
3. Estimates for the interdependencies among the tests; and
4. The "costs" associated with each test (e.g., dollar cost, time required for testing, manpower, training, and resource requirements).

The goal of battery selection is to make use of the given information to formulate a strategy for finding or forming the "best" battery of k tests. The term "best" is used in the sense of trying to find a battery of tests that has good collective performance and low costs, or trying to find the best compromise among the attributes that are important to the decision maker. Determining the value of k (the number of tests in the battery) can be considered as part of the battery selection problem, or a fixed number may already be known before the battery selection process is initiated.

From the preceding description, we can see that the battery selection process would involve the evaluation of multiple objectives. If the number of possible batteries is small (e.g., when the number of tests, n, is relatively small), then we can construct every possible battery of tests and use some simple heuristics to choose from among these batteries. This would require that we know the expected ability of each battery and the costs of each battery, so that we can make trade-offs among the attributes to choose the "best" battery.

In many cases, however, the number of possible choices of batteries may be so large that it becomes computationally burdensome to construct every one and then to try to select the "best" one. In such cases there is a need for computationally efficient procedures and multiple-objective methods that can be used to construct and find a good solution.

In this chapter we discuss several approaches that one can use for battery construction and selection, depending on the problem of interest and the type of information that is available. No matter what approach is used, we must have information about battery performance in order to select a good battery of tests. Thus in Section 4.1 we describe several measures that can be used to summarize the collective performance of a battery of tests. The remainder of this chapter is used to describe three approaches to constructing and selecting good batteries. In Section 4.2 we describe the general process of battery selection when we are able to enumerate all possible batteries of tests. In Section 4.3 we describe a set of heuristics that can be used for constructing combinations of conditionally independent tests that would work well for identifying objects both with and without the property of interest. Once the batteries are constructed, the decision maker can select from among them using the process described in Section 4.2. This approach is most useful if the number of tests (and consequently the number of batteries) is not too large. In Section 4.4 we describe an approach for battery selection that is especially useful when the number of tests is very large. It is an automated method, based on *dynamic programming*, for constructing all of the batteries of conditionally independent tests that are *nondominated*. (See Chapter 2 for a discussion on dynamic programming. A discussion of the concept of nondominated solutions in problems involving multiple objectives is given in Chapter 2 and will also be presented in Section 4.2.) The decision maker can then select from among the nondominated batteries the one that is most preferred.

Each of the methods described in this chapter relies solely on the summary information, such as that which is provided by preliminary analysis of the data base, and the costs of the tests. In many real-world problems,

Battery Selection 127

there may be other considerations that are important as well. For example, the problem may require that certain tests must always be present in the battery, or that one or more tests of a particular type must be present. These types of considerations are problem specific. The reader is referred to Chapter 6 for a specific example of this, where we describe some of the considerations that are involved in selecting batteries of short-term tests for the cancer hazard identification problem and the resulting heuristics that we have found useful.

4.1. COMPUTING MEASURES OF BATTERY PERFORMANCE

In trying to select a "good" battery of tests, we must know how well each battery under consideration is expected to perform. We have already seen that there are a number of commonly used measures for describing the quality and performance of individual tests. These include, for example, the sensitivity, specificity, accuracy, predictivity, assignment potential, and assignment strength of the test (see Section 3.2). Recall that we have chosen to describe the performance of each of the tests by its sensitivity and specificity, because these two measures form natural components in Bayes' theorem (which is the method we use for prediction), and that the existence of interdependencies among the tests can become important when using batteries of tests. In this section we will describe and illustrate several measures for summarizing battery performance that can be computed from the sensitivities and specificities of the tests along with the information about dependencies.

4.1.1. The Sensitivity and Specificity of a Battery of Tests

Let us first see how we can compute the sensitivity of a battery of tests. The sensitivity of a single test is defined as the probability that the test will show a positive result for the property on objects that have the property. If we extend this to multiple tests, we could say that the sensitivity of a battery would be the probability that the *battery will be positive* for the property for objects that have the property. Thus we need to define what is meant by "a battery is positive."

Recall that each test result is given as either a positive or negative result. Thus a battery of test results would be a sequence of positive and negative results. If there were n tests, there would be a total of 2^n possible sequences of positives and negatives. For example, for three tests any of the following eight (2^3) results are possible: $(+_1, +_2, +_3)$, $(+_1, +_2, -_3)$, $(+_1, -_2, +_3)$, $(+_1, -_2, -_3)$, $(-_1, +_2, +_3)$, $(-_1, +_2, -_3)$, $(-_1, -_2, +_3)$, and $(-_1, -_2, -_3)$. The question now is which ones from among the possible results for a given set of tests represent a "positive result for the battery"? Once we have answered this question, then the sensitivity of the battery can be computed by summing the conditional probabilities of all of the sequences that represent positive results for the battery.

To answer the preceding question, we must first define an appropriate "decision rule" for determining when the test results indicate a correct classification. There are a variety of such rules that can be defined. For example, one simple approach that is commonly used would be based on the use of the so-called "positive majority rule." Here the battery is considered positive for the property if the majority of the results are positive for the property. Using the positive majority rule on the example involving three tests, we would compute the sensitivity of the battery as the probability of getting $(+_1, +_2, +_3)$, $(+_1, +_2, -_3)$, $(+_1, -_2, +_3)$, or $(-_1, +_2, +_3)$ on objects that have the property of interest. The sensitivity could thus be computed by the following sum:

$$\Pr(+_1, +_2, +_3 | P) + \Pr(+_1, +_2, -_3 | P)$$
$$+ \Pr(+_1, -_2, +_3 | P) + \Pr(-_1, +_2, +_3 | P)$$

Note that the majority rule requires an odd number of tests in order for a strict majority to be defined.

We can also define the battery to be positive for the property only in the case where all of the tests give positive results. We will call this the "positive consensus rule." If we have n tests, the estimate for the sensitivity of the battery based on the positive consensus rule would be given by the probability $\Pr(+_1, +_2, \ldots, +_n | P)$. If we were considering two different batteries consisting of four tests, for example, and the sensitivities of the two batteries based on the positive consensus rule were given by 0.30 and 0.35, respectively, then we would conclude that the second battery has better performance than the first. This is because there is a greater chance that the second battery will give all positive results on an object with the property of interest than the first battery. We can see that the positive consensus rule can be defined for both an odd and an even number of tests. However, this rule would only be useful in the case where we are comparing batteries with the same number of tests. This is because as we increase the number of tests in a battery, the joint probability of getting all matching results decreases. Thus, for example, when using the positive consensus rule, a battery consisting of 5 tests with sensitivity of 0.3 is not necessarily better than a battery consisting of 10 tests with a sensitivity of 0.25, because in the latter case, the *certainty* of the prediction (i.e., the posterior probability) could be much higher.

A third decision rule for defining when a given battery is positive for the property can be given by examining the Bayesian prediction based on each possible sequence of results for the battery. In particular, we can define the battery to be positive for any sequence of results that produces a prediction that exceeds some prespecified level. If we denote the sequence of test results by $r_1, r_2, \ldots, r_n$ and the level as L, then we say that the battery is positive for every sequence of results that satisfies $\Pr(P|r_1, \ldots, r_n) > L$, where we compute $\Pr(P|r_1, \ldots, r_n)$ by Bayes' formula. The sensitivity of the battery would then be computed by summing the conditional probabilities of getting the results that satisfy the preceding rule. We will call this decision rule the "positive predictive rule." A common level L to use for the positive predictive rule is 0.5 (i.e., any prediction greater than 0.5 would be considered positive for the property). If we were interested only in batteries that would give strong predictions, then the level L could be raised to 0.7 or 0.8. This rule can be used in comparing two batteries of different sizes and can be applied to any number of tests. However, the amount of computation involved can become very extensive when the size of the battery (and consequently the number of possible sequences of results) becomes large. The rule also depends critically on the values of prior probabilities used.

The specificity of a battery of tests can be established in a similar manner. We can define the specificity to be the probability that the *battery gives a negative response* on objects that do not have the property. Decision rules similar to the ones discussed for establishing the positivity of a battery can be defined. For example, we could define the battery to be negative if (1) the majority of the results in the battery are negative—the "negative majority rule," (2) all of the results are negative—the "negative consensus rule," or (3) the Bayesian prediction that the property is absent exceeds a given level [e.g., $\Pr(NP|r_1, \ldots, r_n) > L$]—the "negative predictive rule." The specificity would then be computed by summing conditional probabilities corresponding to the sequences of results that satisfy the chosen rule.

We can now define the computation of the sensitivity and specificity of a battery of tests more formally. Suppose we have a battery of n tests. Denote the set of all 2^n possible sequences of results for the battery by R. For example, for three tests the set R would be defined as $R = \{(+_1, +_2, +_3), (+_1, +_2, -_3), (+_1, -_2, +_3), (+_1, -_2, -_3), (-_1, +_2, +_3), (-_1, +_2, -_3), (-_1, -_2, +_3), (-_1, -_2, -_3)\}$. Let each element of the set R be denoted by R_i ($i = 1, 2, \ldots, 2^n$). Let I^+ denote the set of indices (i.e., subscripts) corresponding to the sequences of responses in the set R that indicate the presence of the property for a given decision rule (e.g., positive majority rule, positive consensus rule, or positive predictive rule); and similarly, let I^- denote the set of indices corresponding to the sequences that indicate the absence of the property for a chosen decision rule. To illustrate the notation more clearly, suppose we have three tests with the set of possible responses, R, defined above. Let R_i ($i = 1, 2, \ldots, 8$) represent the eight responses given in the order presented above (e.g., $R_6 = (-_1, +_2, -_3)$). Thus if we are using the positive majority rule, the set $I^+ = \{1, 2, 3, 5\}$ and for the negative majority rule $I^- = \{4, 6, 7, 8\}$. Similarly, if we are using the positive consensus rule $I^+ = \{1\}$ and if we are using the negative consensus rule $I^- = \{8\}$.

With this notation, the sensitivity and specificity of a battery of tests can be defined as follows:

Sensitivity of a battery:

$$\beta^+ = \sum_{i \in I^+} \Pr(R_i | P) \qquad (4.1.1)$$

Specificity of a battery:

$$\beta^- = \sum_{i \in I^-} \Pr(R_i | NP) \qquad (4.1.2)$$

We can see that once we have chosen the decision rules, the sensitivity and specificity of a battery is simply computed as the sums of the conditional probabilities of the appropriate joint responses of the tests. Each decision rule provides a slightly different measure of battery performance; and the choice of the rule depends on the particular problem. For example, if an odd number of tests is being considered, then the majority rule might be appropriate; if the number of tests is fixed, then a consensus rule might be appropriate; or if the number of tests is not too large, a predictivity rule might be desirable.

The potential difficulty in computing the sensitivity and specificity of a battery using Eqs. (4.1.1) and (4.1.2) is in estimating the joint probabilities: $\Pr(R_i | P)$ and $\Pr(R_i | NP)$. If one has a complete data base containing a large number of responses for the battery on objects both known to have and known not to have the property of interest, then one can estimate the conditional probabilities directly from the data base. In this case, for a given sequence of results, R_i, we can estimate $\Pr(R_i | P)$ from the data base by looking at the objects that have the property and determining the proportion of cases that gave the sequence of results R_i. Similarly, to compute $\Pr(R_i | NP)$ from the data, one can concentrate on the objects that do not have the property, and compute the proportion of cases that gave R_i. If the data base is large (which is actually a desirable property in terms of the accuracy of our results), then these "direct estimates" can require a great deal of computation and, consequently, might be quite impractical to compute. In addition, we may not have a data base containing raw test

Battery Selection

scores, or as is commonly the case, if we do have a data base it may be incomplete or have many gaps. In either of these cases, we may not be able to compute the probabilities directly, and an alternate estimation scheme will be required.

One alternative involves using simply the sensitivities and specificities of the individual tests. If it turns out that the tests in the battery are conditionally independent, then $\Pr(R_i|P)$ and $\Pr(R_i|NP)$ can be computed directly from the sensitivities and specificities of the tests. In particular, if the following results hold for the responses $r_1, r_2, \ldots, r_n$ of tests $T_1, \ldots, T_n$, respectively:

$$\Pr(r_1, r_2, \ldots, r_n|P) = \Pr(r_1|P)\Pr(r_2|P) \cdots \Pr(r_n|P) \qquad (4.1.3)$$

$$\Pr(r_1, r_2, \ldots, r_n|NP) = \Pr(r_1|NP)\Pr(r_2|NP) \cdots \Pr(r_n|NP) \qquad (4.1.4)$$

(i.e., the tests are conditionally independent), then we can see that the right-hand side of each of the preceding expressions can be computed using the sensitivity or specificity of the test. For example, if $r_2 = +$, then $\Pr(r_2|P)$ would be the sensitivity of T_2 and $\Pr(r_2|NP)$ would be one minus the specificity of T_2. Similarly, if r_4 is "$-$," then $\Pr(r_4|P)$ would be one minus the sensitivity of T_4 and $\Pr(r_4|NP)$ would be the specificity of T_4. In this manner we can see that as long as the tests are conditionally independent, the sensitivity and specificity of a battery of tests can be computed directly from the sensitivities and specificities of the individual tests, regardless of the decision rule that is used.

In order to check whether Eqs. (4.1.3) and (4.1.4) hold for a particular battery, we again need a complete data base. One way of getting around this problem is to utilize the results based on pairs of tests given by the preliminary analysis (see Section 3.4). We have found that if each pair of tests in the battery is conditionally independent and as long as the number of tests in the battery is not too large (say, 10 or less), then the estimates given in Eqs. (4.1.3) and (4.1.4) remain fairly accurate (Pet-Edwards, 1986).

If the tests are found to be conditionally dependent, then Eqs. (4.1.3) and (4.1.4) should not be used. To account for the dependencies, we can adjust the two equations. In particular, they can be adjusted as follows:

$$\Pr(r_1, r_2, \ldots, r_n|P) = K_P(R_i)\Pr(r_1|P)\Pr(r_2|P) \cdots \Pr(r_n|P) \qquad (4.1.5)$$

$$\Pr(r_1, r_2, \ldots, r_n|NP)$$
$$= K_{NP}(R_i)\Pr(r_1|NP)\Pr(r_2|NP) \cdots \Pr(r_n|NP) \qquad (4.1.6)$$

where $K_P(R_i)$ is the adjustment for dependence for the sequence of results R_i in population P and $K_{NP}(R_i)$ is the adjustment for dependence for the sequence of results R_i in population NP. To compute the actual magnitudes of $K_P(R_i)$ and $K_{NP}(R_i)$ would again require a complete data base. For example, $K_P(R_i)$ would be computed as the following ratio:

$$K_P(R_i) = \frac{\Pr(r_1, \ldots, r_n | P)}{\Pr(r_1 | P) \cdots \Pr(r_n | P)}$$

We can approximate these two dependence factors from the pairwise conditional dependence information. We have found that an estimate based on the product of the n-minus-one pairwise K_P measures (for a description of the K_P measures, see Section 3.4) whose absolute deviations from the value one are the greatest provides a fairly accurate estimate as long as the number of tests is not too large (Pet-Edwards, 1986).

To illustrate how the computation of the sensitivity of a battery of tests would proceed, the following hypothetical example is given. (Note that the computation of the specificity of a battery of tests would be done in an analogous fashion.) In order to keep the illustration as simple as possible, suppose that we have three tests and we would like to estimate the sensitivity of the battery of tests for the case in which we use the positive majority rule. Using Eq. (4.1.1) we see that the sensitivity of the battery would be given as

$$\beta^+ = \Pr(+_1, +_2, +_3 | P) + \Pr(+_1, +_2, -_3 | P)$$

$$+ \Pr(+_2, -_2, +_3 | P) + \Pr(-_1, +_2, +_3 | P)$$

Suppose we know that the sensitivities of the three tests, T_1, T_2, and T_3, are 0.7, 0.8, and 0.9, respectively. If we knew that the three tests were conditionally independent, or as an estimate that the tests when taken as pairs were conditionally independent, then we could use Eq. (4.1.3) to estimate each of the joint probabilities given in the preceding sum. In particular, the sensitivity of the battery would be given by

$$\beta^+ = (0.7)(0.8)(0.9) + (0.7)(0.8)(0.1) + (0.7)(0.2)(0.9) + (0.3)(0.8)(0.9)$$

$$= 0.902$$

Suppose, however, that we found that these tests were not conditionally independent and that preliminary analysis of our data base gave us the following results for the pairs of tests: $K_P(+_1, +_2) = 1.0$, $K_P(+_1, +_3) = 1.05$,

Battery Selection 133

and $K_P(+_2, +_3) = 1.1$. In order to compute the sensitivity of the battery, we would compute the joint conditional probabilities using Eq. (4.1.5). This would require estimates for $K_P(R_i)$ for the four sequences of responses, R_i ($i \in I^+$) comprising the sensitivity of the battery [i.e., $K_P(+_1, +_2, +_3)$, $K_P(+_1, +_2, -_3)$, $K_P(+_1, -_2, +_3)$, and $K_P(-_1, +_2, +_3)$]. We can approximate these dependence factors using the pairwise information. In particular, consider the estimate for $K_P(+_1, +_2, +_3)$. Here we would select the two pairwise estimates [note that $(n - 1) = 2$] that deviate the most from one and take their product. Note that the K_P measures for the pairs of tests (T_1, T_3) and (T_2, T_3) deviate the most from one. We see that these two estimates would be $K_P(+_1, +_3)$ and $K_P(+_2, +_3)$, and our estimate for $K_P(+_1, +_2, +_3)$ would be given by the product of 1.05 and 1.1, namely, 1.155. Using these same test pairs we would compute the remaining estimates as follows:

$$K_P(+_1, +_2, -_3) = K_P(+_1, -_3)K_P(+_2, -_3)$$

$$K_P(+_1, -_2, +_3) = K_P(+_1, +_3)K_P(-_2, +_3)$$

$$K_P(-_1, +_2, +_3) = K_P(-_1, +_3)K_P(+_2, +_3)$$

In order to compute $K_P(+_i, -_j)$ and $K_P(-_i, +_j)$ ($i = 1, 2$ and $j = 3$), we use the relations described in Section 3.4 given in Eqs. (3.4.11) and (3.4.12). For example, $K_P(+_1, -_3)$ would be given by $[1 - K_P(+_1, +_3)\alpha_3^+)]/(1 - \alpha_3^+)$ which is $[1 - (1.05)(0.9)]/0.1 = 0.55$. Similarly, $K_P(+_2, -_3) = 0.1$, $K_P(-_2, +_3) = 0.6$, and $K_P(-_1, +_3) = 0.883$. Thus $K_P(+_1, +_2, -_3) = 0.055$, $K_P(+_1, -_2, +_3) = 0.63$, and $K_P(-_1, +_2, +_3) = 0.971$.

The estimate for the sensitivity of this battery of dependent tests would be given by

$$\beta^+ = 1.155(0.7)(0.8)(0.9) + 0.055(0.7)(0.8)(0.1) + 0.63(0.7)(0.2)(0.9)$$

$$+ 0.971(0.3)(0.8)(0.9)$$

$$= 0.831$$

(Note in this example that the positive dependencies between the tests have degraded the sensitivity of the battery from 0.902 to 0.831.)

4.1.2. Battery Performance Measures Based on Predictivities

We can also compute measures of battery performance based on the magnitudes of the Bayesian predictions given by particular sequences of test results. These are the so-called "predictivity-based" performance measures. In order to see how predictivities can be used to compare tests and how they can be computed for batteries of tests, we will begin by looking at the predictivities of individual tests. Recall that Bayes' formula for a single positive test result would give us the probability that the property is present given that the test result was positive; and for a negative result could give us the probability that the property is absent given that the test was negative. Suppose that we would like to compare the following three tests based on their predictivities:

Test	Sensitivity	Specificity
T_1	0.8	0.8
T_2	0.9	0.6
T_3	0.6	0.9

Suppose that before testing, we believe there is a 50-50 chance that the property is present in a set of objects [i.e., $\Pr(P) = \Pr(NP) = 0.5$] and we would like to see how strongly a positive result would indicate the presence of the property and how strongly a negative result would indicate the absence of the property for each of the tests. We could determine this by computing the Bayesian prediction using a prior probability of 0.5. For example, for the second test the computations of the predictivities would proceed as follows:

$$\Pr(P|+_2) = \frac{(0.5)(0.9)}{(0.5)(0.9) + (0.5)(0.4)} = 0.69$$

$$\Pr(NP|-_2) = \frac{(0.5)(0.6)}{(0.5)(0.6) + (0.5)(0.1)} = 0.86$$

The computations for the remaining two tests would proceed in an analogous fashion. The results for the three tests are summarized in Table 4.1.1.

We may want to use the predictivities in order to determine which one of the three tests to use in a particular application. For example, suppose that according to company regulations, any chemical whose potential for

TABLE 4.1.1. Example of the Predictivities of Individual Tests

Test	Sensitivity	Specificity	$\Pr(P\|+)$	$\Pr(NP\|-)$
T_1	0.8	0.8	0.8	0.8
T_2	0.9	0.6	0.69	0.86
T_3	0.6	0.9	0.86	0.69

Battery Selection

carcinogenicity exceeds 0.75 is considered hazardous and therefore undesirable. If the three preceding tests were indicators of potential carcinogenicity, and the company wanted to select one of the three tests to use, the only ones that would give conclusive results would be T_1 [Pr(P|+) = 0.8] and T_3 [Pr(P|+) = 0.86]. Since T_3 is more conclusive, the company would probably select that test. Alternatively, if the regulation was that any substance is considered safe if its potential for carcinogenicity is below 0.25 and is dangerous if its potential for carcinogenicity exceeds 0.75, then only T_1 would give conclusive results in both cases [i.e., Pr(P|+) = 0.8 > 0.75 and Pr(P|−) = 1 − Pr(NP|−) = 0.2 < 0.25].

If the prior probability is not known, then we can compute the predictivities as a function over the range of prior probabilities. We can then graph the predictivity as a function of the prior probabilities, thus obtaining graphs similar to those given in Fig. 3.4.1. Examples of this are given in Chapter 6.

We can compute predictivities to compare batteries of tests as well. The predictivities for a battery of tests would be given as the probability that the property is present given that the *battery is positive*—the "positive predictivity" denoted by θ^+, and the probability that the property is absent given that the *battery is negative*—the "negative predictivity" denoted by θ^-. As in the case of the sensitivity and specificity of a battery of tests, various measures of battery predictivity can be defined based on different decision rules for determining when a battery is positive and when a battery is negative.

For example, we can employ (1) the majority rule, (2) the consensus rule, or (3) the predictive rule as described earlier. Once we have selected a rule, the computation of the positive and negative predictivities of the battery would follow directly from Bayes' formula. In particular, suppose we would like to determine the probability that the property is present given that the battery is positive (i.e., θ^+). Let us denote a positive battery by "+B." Thus we would like to determine Pr(P|+B). Using Bayes' formula, we can rewrite this as

$$\theta^+ = \Pr(P|+B) = \frac{1}{1 + \frac{\Pr(NP)\Pr(+B|NP)}{\Pr(P)\Pr(+B|P)}} \quad (4.1.7)$$

Similarly, the predictivity for a negative battery, "−B," would be given as

$$\theta^- = \Pr(NP|-B) = \frac{1}{1 + \frac{\Pr(P)\Pr(-B|P)}{\Pr(NP)\Pr(-B|NP)}} \quad (4.1.8)$$

Note that in Eq. (4.1.7), for example, we need to know the probability that the battery is positive given that we are in population P. This would simply be the sensitivity of the battery using the chosen decision rule. Similarly, $\Pr(+B|NP)$ would be one minus the specificity of the battery, $\Pr(-B|P)$ would be one minus the sensitivity of the battery, and $\Pr(-B|NP)$ would be the specificity of the battery. Each of these probabilities for a particular decision rule would thus be computed in the fashion described earlier. The following are illustrations of the computations involved.

Suppose we have the three tests given in Table 4.1.1. These three tests when used in a battery can result in eight possible sequences of results. If the tests are conditionally independent, then we can compute the probabilities that each of these sequences would occur in a particular population by using Eqs. (4.1.3) and (4.1.4). The results are displayed below:

Sequence	T_1	T_2	T_3	$\Pr(R_i|P)$	$\Pr(R_i|NP)$
R_1	+	+	+	0.432	0.008
R_2	+	+	−	0.048	0.072
R_3	+	−	+	0.288	0.012
R_4	+	−	−	0.032	0.108
R_5	−	+	+	0.108	0.032
R_6	−	+	−	0.012	0.288
R_7	−	−	+	0.072	0.048
R_8	−	−	−	0.008	0.432

Suppose we are using the majority rule for deciding when the battery is positive and when the battery is negative. If we would like to determine the probability that the property is present given that the battery is positive, we would use Eq. (4.1.7). Note that $\Pr(+B|P)$ would be given by

$$\Pr(R_1|P) + \Pr(R_2|P) + \Pr(R_3|P) + \Pr(R_5|P) = 0.876$$

and that $\Pr(+B|NP)$ would be given by

$$\Pr(R_1|NP) + \Pr(R_2|NP) + \Pr(R_3|NP) + \Pr(R_5|NP) = 0.124$$

Thus using a prior probability of 0.5 and using Eq. (4.1.7), we would find that the predictivity of the battery would be

$$\Pr(P|+B) = 1/[1 + (0.124/0.876)] = 0.876$$

Notice that this is a higher predictivity than one could achieve by using any of the tests alone (refer to Table 4.1.1). We can find the negative predictivity, $\Pr(NP|-B)$, in an analogous fashion.

Battery Selection

Suppose we are using the same battery of tests (where the tests are conditionally independent), but are now interested in seeing the predictivity based on the consensus rule. If we would like to determine $\Pr(\text{NP}|-B)$ using Eq. (4.1.8), then $\Pr(-B|\text{NP})$ would be given by $\Pr(R_8|\text{NP}) = 0.432$ and $\Pr(-B|\text{P})$ would be given by $\Pr(R_8|\text{P}) = 0.008$. If the prior probability is 0.5, then the predictivity, $\Pr(\text{NP}|-B)$, would be 0.844.

Suppose instead that the three tests were not conditionally independent, but rather that the following results were obtained from preliminary analysis:

$K_P(\cdot)$	$K_{NP}(\cdot)$
$K_P(+_1, +_2) = 1.0$	$K_{NP}(-_1, -_2) = 1.2$
$K_P(+_1, +_3) = 1.0$	$K_{NP}(-_1, -_3) = 1.05$
$K_P(+_2, +_3) = 1.0$	$K_{NP}(-_2, -_3) = 1.1$

We see that in population P the tests are pairwise independent. Thus as an approximation of $\Pr(R_8|\text{P})$ we can use Eq. (4.1.3), which would give us 0.008 (the same value as before). In order to compute $\Pr(R_8|\text{NP})$ we would need to use Eq. (4.1.6). An estimate for $K_{NP}(R_8)$ would be given by the product of the two pairwise dependence measures that differ the most from one. In particular, an estimate for $K_{NP}(R_8)$ would be given by $(1.2)(1.1) = 1.32$ and the estimate for $\Pr(-B|\text{NP})$ would then be given by $(1.32)(0.432) = 0.570$. We can now compute the value of the predictivity (assuming the prior probability is 0.5) using Eq. (4.1.8): $1/[1 + (0.008/0.570)] = 0.986$.

The sensitivities, specificities, and predictivities for various decision rules provide a wide variety of measures for evaluating battery performance. The choice of which measure or set of measures to use will depend on the specific problem. It will depend, for example, on (1) whether the choice of batteries is large (where we might want to choose decision rules that would require fewer computations), (2) whether we are trying to minimize false negatives or positives (where battery sensitivity or specificity, respectively, might be appropriate), or (3) whether we are searching for batteries that would give us conclusive results (where predictivities might be appropriate).

When the number of choices of batteries is very large, the amount of computation involved in computing any of the preceding battery performance measures can be prohibitive. In this case it would be useful to use some heuristics or some methodology that would allow us to construct only a subset of the possible batteries (e.g., the ones that would "perform" best). In this chapter, we describe three such methods. The "mechanics" of these methods, which we describe in Sections 4.2-4.4, can limit the form that the performance measures can take. In the following section, we describe how battery selection would proceed in the case where we can enumerate every possible battery option.

4.2. BATTERY SELECTION USING ENUMERATION

In the Introduction to this chapter we described the problem of selecting a battery of tests and noted that it involves making trade-offs between the performances of the batteries and various measures of battery cost. In Section 4.1 we described and illustrated how we can compute several useful measures of battery performance. We have not yet described how to compute the costs. In general it is simpler to devise measures of the costs of a particular battery than it is to compute its performance. For example, the dollar cost of performing a battery of tests might be given by the sum of the costs of the individual tests, the "time cost" of a given battery might be given by the maximum of the times required by each test if the tests are performed simultaneously or the sum of the times if the tests are performed in sequence, and the "resource cost" might be given by the sum of the resources required by each test in the battery. Thus, for any set of tests, it would be possible to compute the performance (using the methods of the preceding section) and the "costs" for every possible combination of tests; however, when the number of tests is large it may be impractical to do this.

If we have n tests, then there would be a total of $(2^n - 1)$ possible batteries. For example, if we had 10 tests then we could construct 1023 possible batteries, if we had 20 tests then we could construct 1,048,575 possible batteries, and if we had 50 tests then we could construct 1.12×10^{15} batteries. The latter cases would require a great deal of computation and would necessitate the use of more efficient methods. Two of these methods are described in Sections 4.3 and 4.4. For the remainder of this section, we will assume that we have reasonable (i.e., relatively small) number of tests so that all of the combinations can be constructed and the attributes of each battery (i.e., the performances and the costs) can be computed with a reasonable amount of effort.

Suppose that we have n tests, that we have enumerated every possible combination of these tests, and that we have computed appropriate performance measures (the appropriateness determined by the particular problem) and the costs for each battery. For example, let B_i represent battery i ($i = 1, 2, \ldots, 2^n - 1$), let β_i^+ be the computed sensitivity of battery i using the positive majority rule, let β_i^- be the computed specificity of battery i using the negative majority rule, and let C_i be the dollar cost of performing battery i determined by the sum of the individual costs of the tests in the battery. As a concrete example, suppose we have a data base containing

Battery Selection

TABLE 4.2.1. Results of Preliminary Analysis on Five Hypothetical Tests where All Test Pairs Are Conditionally Independent

Test	Sensitivities	Specificities	Costs ($)
T_1	0.8	0.6	500
T_2	0.8	0.55	600
T_3	0.6	0.88	1000
T_4	0.6	0.52	100
T_5	0.55	0.505	200

five hypothetical tests—T_1, T_2, T_3, T_4, and T_5—and that the costs of the tests and the results of preliminary analysis of the data base are summarized in Table 4.2.1. We can construct 31 (i.e., $2^5 - 1$) possible batteries. However, since we are using the majority rule for computing the sensitivities and specificities, only batteries composed of an odd number of tests will be used. These batteries along with their costs and their sensitivities and specificities (computed using the methods of Section 4.1) are summarized in Table 4.2.2.

The problem of selecting the "best" battery from among the options listed in Table 4.2.2 is a multiple-objective problem. Recall from Chapter

TABLE 4.2.2. All Possible Batteries of Tests Constructed from Table 4.2.1

Battery	(B_i)	Sensitivities (β_i^+)	Specificities (β_i^-)	Costs (C_i) ($)
T_1	B_1	0.8	0.6	500
T_2	B_2	0.8	0.55	600
T_3	B_3	0.6	0.88	1000
T_4	B_4	0.6	0.52	100
T_5	B_5	0.55	0.505	200
T_1, T_2, T_3	B_6	0.83	0.76	2100
T_1, T_2, T_4	B_7	0.83	0.58	1200
T_1, T_2, T_5	B_8	0.82	0.577	1300
T_1, T_3, T_4	B_9	0.744	0.748	1600
T_1, T_3, T_5	B_{10}	0.722	0.742	1700
T_1, T_4, T_5	B_{11}	0.722	0.562	800
T_2, T_3, T_4	B_{12}	0.744	0.724	1700
T_2, T_3, T_5	B_{13}	0.722	0.717	1800
T_2, T_4, T_5	B_{14}	0.722	0.537	900
T_3, T_4, T_5	B_{15}	0.624	0.702	1300
T_1, T_2, T_3, T_4, T_5	B_{16}	0.801	0.705	2400

2, Section 2.3, that an optimal solution may not exist when we are "solving" multiple-objective problems. Consider the results given in Table 4.2.2 and suppose that we are trying to find a battery that would have maximum sensitivity, maximum specificity, and minimum cost. Note that such an "optimal solution" does not exist. Instead we have some batteries that are good in one or two attributes and not so good in the other(s). Recall also that we can eliminate *dominated solutions* from consideration, since a dominated solution would never be preferred by a rational decision maker. A solution is dominated (inferior, or not Pareto-optimal) if there exists another option that is better (or at least as good) in each of the attributes. Nine of the battery options listed in Table 4.2.2 are dominated by other batteries. For example, consider Battery 16. Note that its sensitivity is 0.801, its specificity is 0.705 and its cost is \$2400. Now consider Battery 6, with a sensitivity, specificity, and cost of 0.83, 0.76, and \$2100, respectively. Note that Battery 6 has a higher sensitivity, a higher specificity, and lower cost than Battery 16. Thus Battery 16 is dominated by Battery 6 and we can eliminate Battery 16 from consideration. Similarly, we can see that Battery 1 dominates Batteries 2, 14, and 11; Battery 4 dominates Battery 5; Battery 7 dominates Battery 8; and Battery 9 dominates Batteries 10, 12, and 13. The seven remaining nondominated batteries are listed in Table 4.2.3. It is now up to the decision maker to choose his/her most preferred solution from among these nondominated options.

The choice of which of the nondominated batteries listed in Table 4.2.3 is preferred will depend on the specific problem under consideration and the preferences of the decision maker. For example, suppose that the decision maker is indifferent about the cost of the battery, but that he/she would like to minimize the number of false positives and false negatives. In this case note that only the two batteries listed in Table 4.2.4 are nondominated. The decision maker must now decide whether minimizing false negatives (i.e., maximizing the sensitivity) is more important or whether

TABLE 4.2.3. All Nondominated Batteries from Table 4.2.2

Battery	(B_i)	Sensitivities (β_i^+)	Specificities (β_i^-)	Costs (C_i) (\$)
T_1	B_1	0.8	0.6	500
T_3	B_3	0.6	0.88	1000
T_4	B_4	0.6	0.52	100
T_1, T_2, T_3	B_6	0.83	0.76	2100
T_1, T_2, T_4	B_7	0.83	0.58	1200
T_1, T_3, T_4	B_9	0.744	0.748	1600
T_3, T_4, T_5	B_{15}	0.624	0.702	1300

Battery Selection

TABLE 4.2.4. All Nondominated Batteries from Table 4.2.3 when Only the Sensitivities and Specificities Are Considered

Battery	(B_i)	Sensitivities (β_i^+)	Specificities (β_i^-)
T_3	B_3	0.6	0.88
T_1, T_2, T_3	B_6	0.83	0.76

minimizing false positives (i.e., maximizing the specificity) is more important. In other words, since there does not exist a single battery that is best in both attributes, the decision maker must make trade-offs between the two objectives. The choice will most certainly depend on the problem under consideration. For example, if the decision maker is the manager of a chemical company and the tests are used for identifying potential cancer hazards, the manager might want to have a battery that would minimize false negatives since it could be quite disastrous for the company to continue to develop a chemical that is later found to be a cancer hazard. In this case Battery 6, composed of the three tests, T_1, T_2, and T_3, might be preferred. On the other hand, a regulatory agency might want to select the battery that would minimize false positives since it would want to begin by regulating chemicals most likely to be hazardous. In this case Battery 3, composed of the single test T_3, might be preferred.

As a second illustration, suppose that the decision maker is the manager of a chemical company and that he/she would like to minimize false negatives, to minimize costs, and is indifferent to false positives. In this case the three options listed in Table 4.2.5 are nondominated. We can graph the false-negative rate (i.e., one minus the sensitivity) versus the cost for each of the nondominated battery options (see Fig. 4.2.1). Again we can see that there is no single optimal solution, and the decision maker must make trade-offs between the cost and the false-negative rate of the battery.

TABLE 4.2.5. All Nondominated Batteries from Table 4.2.3 when Only the Sensitivities and Costs Are Considered

Battery	(B_i)	Sensitivities (β_i^+)	Costs (C_i) ($)
T_1	B_1	0.8	500
T_4	B_4	0.6	100
T_1, T_2, T_4	B_7	0.83	1200

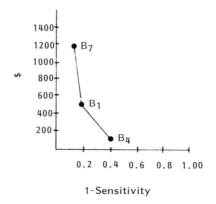

FIGURE 4.2.1. Graph of false negatives versus cost for data given in Table 4.2.5.

Refer to Fig. 4.2.1 and note that if we move from Battery 7 to Battery 1, we decrease the cost by $700 while decreasing the performance of the battery by only 0.03. However, when we move from Battery 1 to Battery 4, we decrease the cost by $400.00 while decreasing the performance by 0.2. The manager might be willing to make the first trade-off, while it is likely that he/she would be unwilling to make the second trade-off. Figures of the type given in Fig. 4.2.1 can assist the decision maker in making trade-offs between the various options.

Batteries composed of the five tests given in Table 4.2.1 could also be analyzed in terms of sensitivities and specificities determined by consensus rules or predictivity rules or, if the decision maker wants to evaluate the batteries in terms of how decisive they are, then predictivity measures based on various decision rules might be used. The battery options can then be evaluated using the preceding approach.

The approach described in this section can always be applied, regardless of the performance measures used or the costs. However, when the number of battery options is large, the amount of computation required to compute the performance of every battery option can be very extensive. The remaining sections of this chapter describe approaches that might be useful in these cases.

4.3. A HEURISTIC APPROACH FOR CONSTRUCTING BATTERIES WITH GOOD PERFORMANCE

As described in the preceding section, it may be very inconvenient to enumerate the performances of every possible battery of tests when the

Battery Selection

number of batteries is large. In this case, if we could develop a strategy for constructing only the batteries that would have the "best" performance based on information on the individual tests, then we might be able to eliminate some of the work involved in computing the performances of the batteries. The types of information that are available from preliminary analysis are the sensitivities and specificities of the tests and the interdependencies between pairs of tests. We can utilize this information to determine the performances of the batteries and can develop some general guidelines for constructing a subset of batteries with good performance.

We have found in general that when tests are positively dependent, the performance of a battery of such tests would not be as good as a battery composed of tests with the same sensitivities and specificities that are conditionally independent. Thus, if we have an adequate number of conditionally independent tests, then we can devise a strategy (heuristic) for constructing the best batteries from the sensitivities and specificities of the individual tests. In order to see how this can be done, we will examine how tests with various levels of performance would work.

As illustrated earlier (and will be further illustrated in Chapter 5), Bayes' formula allows us to calculate the prediction of a battery of conditionally independent tests based simply on the knowledge of the sensitivities and specificities of the individual tests along with a prior probability that the property is present. Note that a test with a high sensitivity and a high specificity would be able to detect both objects with and without the property of interest with a high degree of accuracy. A test with high sensitivity but with low-to-moderate specificity would be highly selective for objects with the property but not for the objects without the property. This means that this type of test would likely give a positive result when it encounters an object with the property, but would give positive or negative results when it encounters objects without the property. Thus if we were to use such a test to predict the presence of the property, a positive response could occur from an object with the property (a "true positive") or from a misclassified object that does not have the property (a "false positive") but a negative response would occur most often only for objects without the property. Thus a test with a high sensitivity and a low specificity would be a good predictor of objects without the property, but not so good for objects with the property.

In a similar manner we can see that a test with low-to-moderate sensitivity and high specificity would be a good predictor of objects with the property, but a poor-to-moderate predictor of objects without the property. And finally, a test that has a low sensitivity and a low specificity would be a poor predictor of either sets of objects. Thus we can see that tests can be divided into four classes—Class I: tests with high sensitivities and high specificities; Class II: tests with high sensitivities and low specificities; Class III: tests with low sensitivities and high specificities; and Class IV: tests with low sensitivities and low specificities (see Table 4.3.1). The definitions of what is meant by the terms "high" and "low" are arbitrary and would be determined by the particular problem. For example, in the cancer hazard identification problem, where the tests are short-term *in vitro* or *in vivo* assays for genotoxicity, we have found that the range of values for the sensitivities and specificities of the assays is approximately 0.65–0.90 (see McCann *et al.*, 1975; and other references cited in Heinze and Poulson, 1983). An arbitrary level of 0.75 was chosen since it was approximately midpoint in the range and it allowed the tests to be divided up into approximately equally sized groups. Once we have chosen the cutoff level for the sensitivity and specificity, note that the Class II and Class III tests are dominated by the Class I tests and that the Class IV tests are dominated by the Class I, Class II, and Class III tests.

Inspection of Table 4.3.1 reveals that a Class II test would be a good predictive test for objects without the property; however, it would not perform as well in predicting objects with the property if the test result was positive. The reverse would be true for a test from Class III. It would be logical to expect that if a test from Class II is coupled with one from Class III, then they would complement one another. The strong points of each test can thus be effectively utilized. Agreement between the two tests (whether positive or negative) would further strengthen our confidence in the prediction. However, disagreement between the test results would produce a "null effect" on our state of knowledge about the property of the

TABLE 4.3.1. Scheme for Classifying the Tests Based on Their Performances

Use of test	Classes of tests[a]			
	Class I $L < \alpha^+ \leq 1$ $L < \alpha^- \leq 1$	Class II $L < \alpha^+ \leq 1$ $0 \leq \alpha^- \leq L$	Class III $0 \leq \alpha^+ \leq L$ $L < \alpha^- \leq 1$	Class IV $0 \leq \alpha^+ \leq L$ $0 \leq \alpha^- \leq L$
Detecting P	Moderate to good	Moderate to good	Poor to moderate	Poor to moderate
Detecting NP	Moderate to good	Poor to moderate	Moderate to good	Poor to moderate
Predicting P	Moderate to good	Poor to moderate	Moderate to good	Poor to moderate
Predicting NP	Moderate to good	Moderate to good	Poor to moderate	Poor to moderate

[a] L represents an arbitrary "cutoff" point determined by the particular application.

Battery Selection

tested object. This can be verified by examining the Bayesian predictions for such combinations of tests.

The preceding observations have led us to the following strategy for constructing batteries with "good performance."

4.3.1. Battery Selection CPBS Strategy for Constructing Batteries of Tests

Begin with the list of sensitivities, specificities, and dependencies given by preliminary analysis.

Step 1. Select an appropriate classification level for the sensitivities and specificities in order to divide the tests in the four previously described classes.

Step 2. Divide the tests into the four classes.

Step 3. Construct batteries using the following rules:
  (a) An odd number of tests should be used if majority rule is going to be used as a decision rule for computing the performance of the battery.
  (b) Only combine tests if they are conditionally independent.
  (c) Use as many Class I tests as are available.
  (d) If the objective is to minimize false nagatives or to have strong predictions for objects without the property of interest and the decision maker is indifferent about false positives, then use tests from Class II.
  (e) If the objective is to minimize false positives or to have strong predictions for objects with the property of interest and the decision maker is indifferent about false negatives, then use tests from Class III.
  (f) If minimizing false negatives and false positives are both important, then always couple a Class II test with a Class III test.
  (g) Do not utilize Class IV tests.

If both the sensitivity and specificity of the battery are important (i.e., both false positives and false negatives are important) and a decision rule based on majority rule is going to be used to compare the batteries, then a typical battery formed by using the above strategy would consist of an odd number of tests from Class I plus some number (which may be zero) of tests from Class II and an equal number from Class III. If the predictivity of the battery based on the consensus rule is going to be used and the decision maker is only interested in getting strong predictions for objects with the property of interest, then a typical battery would consist of tests from Class I and any number (which may be zero) of tests from Class III.

By using the preceding strategy, we can eliminate some of the combinations from consideration. This preliminary screening of all the possible batteries can reduce the amount of computation involved. The following example from the cancer hazard identification problem adapted from Pet-Edwards *et al.* (1985) is used to illustrate the procedure.

EXAMPLE 4.3.1

Consider the nine short-term *in vitro* tests for carcinogenicity displayed below. Preliminary analysis of the Gene-Tox data base gives the sensitivities and specificities summarized in the following table:

Test	Sensitivity	Specificity
DRL	0.803	0.80
HMA	0.757	0.99
BHK	0.781	0.80
PrM	0.889	0.50
SHE	0.906	0.67
VET	0.890	0.44
Sty	0.612	0.81
EcW	0.633	0.86
UDS	0.560	0.99

The chi-square statistic indicates that the following test pairs are likely to be dependent:

Sty, Ecw
Sty, DRL
Sty, UDS
DRL, UDS
EcW, DRL
EcW, UDS

In addition, cluster analysis using the single-link method indicates that the following test pairs are likely to be positively dependent:

PrM, EcW
SHE, BHK
HMA, SCE
Cvt, SCE
Sty, EcW

Battery Selection

Note that if we use a cutoff of 0.75 for the sensitivities and specificities, we can divide the nine tests into four classes as follows:

>Class I: DRL, HMA, BHK
>Class II: PrM, SHE, VET
>Class III: Sty, EcW, UDS
>Class IV: —

If we were to construct all possible batteries of these nine tests, then there would be $2^9 - 1 = 511$ possible batteries to consider. Suppose, however, that we plan to compare the batteries based on their sensitivities, specificities, and predictivities, each computed based on the majority rule. If we utilize the strategy for constructing batteries of tests we would see that each battery would consist of one or three tests from Class I (i.e., an odd number), some number (possibly zero) from Class II and an equal number from Class III, and no test pairs that are suspected to be statistically dependent. Note that the tests in Class I are not suspected to be statistically dependent, so each one or all three may be used in a battery. Now consider one of the Class I tests, say DRL. Note that DRL is statistically dependent on each of the Class III tests (see the results of the chi-square statistic). Thus the test, DRL, cannot be combined with a test from Class III and consequently not from Class II either. If we continue with this "brute-force" approach to constructing batteries, we find that only 17 out of the 511 possible batteries would satisfy the preceding rules. These batteries along with their computed sensitivities, specificities, and predictivities are summarized in Table 4.3.2.

The preceding example illustrated how the battery construction strategy can screen the possible batteries for only those that would work best for the particular application. This can result in a considerable reduction in the number of batteries to consider. At this point, other factors such as the cost of each test, the time it takes to test the objects, the number of personnel it takes, etc., could be taken into account along with the battery performance in order to make the selection of the most preferred battery. The process described in Section 4.2 would enable the decision maker to accomplish this.

Although the preceding strategy for battery construction can be quite useful for reducing the number of battery choices, it may turn out that (1) the number of tests in one or more of the classes is large, (2) all of the tests fall in one class, or (3) test pairs cannot be eliminated through considerations of dependencies. In any of the preceding cases, the strategy for battery construction outlined above could result in a potentially large number of batteries to consider, and consequently the use of a more efficient procedure might be useful. The method described in the following section is designed for this purpose.

TABLE 4.3.2. Possible Batteries Using the Battery Construction Strategy[a]

Battery			Sensitivity $(\beta+)$	Specificity $(\beta-)$	Predictivity $(\theta+)$	Predictivity $(\theta-)$
	DRL		0.803	0.800	0.801	0.802
	BHK		0.781	0.800	0.796	0.785
	HMA		0.757	1.00	1.00	0.805
DRL	BHK	HMA	0.877	0.960	0.956	0.886
BHK	PrM	Sty	0.867	0.803	0.815	0.858
BHK	VET	Sty	0.867	0.785	0.801	0.855
BHK	VET	EcW	0.873	0.812	0.823	0.865
BHK	PrM	UDS	0.852	0.900	0.895	0.859
BHK	VET	UDS	0.852	0.889	0.885	0.857
HMA	SHE	Sty	0.864	0.935	0.930	0.873
HMA	PrM	Sty	0.857	0.903	0.898	0.863
HMA	VET	Sty	0.857	0.892	0.888	0.862
HMA	SHE	EcW	0.870	0.952	0.948	0.880
HMA	VET	EcW	0.863	0.920	0.915	0.870
HMA	SHE	UDS	0.849	1.00	1.00	0.869
HMA	PrM	UDS	0.841	1.00	1.00	0.863
HMA	VET	UDS	0.841	1.00	1.00	0.863

[a] Taken from Pet-Edwards et al. (1985).

4.4. A STRATEGY FOR CONSTRUCTING BATTERIES USING DYNAMIC PROGRAMMING

When the number of tests is very large and the approach of the preceding section also fails to substantially reduce the amount of computation, then an approach based on dynamic programming (see Section 2.4) might be useful. In this section, we will describe how dynamic programming may be used to generate nondominated batteries of tests for the case where the tests are conditionally independent. We noted earlier that dynamic programming has been used successfully in problems that can be posed as sequential (multistage) problems that satisfy Bellman's "principle of optimality" (Bellman and Dreyfus, 1962):

> An optimal policy has the property that whatever the initial state and decisions are, the remaining decisions must constitute an optimal policy with regard to the state resulting from the first decision.

In order to use the concept of dynamic programming for constructing batteries of tests, we must be able to structure the construction of batteries as a sequential process. One way of doing this is to define the stage of the dynamic program (see Section 2.4 for clarification of the term "stage") as the maximum number of tests in the battery. Then the construction of batteries can be given sequentially as follows: we construct new batteries (i.e., batteries at Stage m) from old batteries (i.e., batteries at Stage $m - 1$) by adding a new test to the old batteries. In order for this process of construction to satisfy Bellman's principle, the information given at a previous stage must be all that is necessary to compute the batteries at the next stage, and this process must lead to the optimal solution.

The description of dynamic programming given in Section 2.4 was for the single-objective case. The battery selection problem, however, is a multiple-objective problem. Dynamic programming can be extended to handle multiple objectives. We will not give the theoretical details of this formulation here [the reader is referred to Yu (1985), Tauxe *et al.* (1979), and Pet-Edwards (1986)], but will discuss some of the technicalities of the use of multiple-objective dynamic programming that lead to restrictions on the functional forms of the performance measures.

For our purposes, we would like the dynamic-programming-based battery construction process to construct all of the nondominated batteries. For simplicity, suppose we are dealing with a problem where only two objectives are important. For clarity of presentation, suppose these objectives are to minimize the cost and to maximize one of the measures of performance described in Section 4.1. Consider the following sequential battery construction process involving n tests:

Stage 1. List all of the nondominated tests.

Stage 2. To each of the tests in Stage 1 add a test and check whether the "new battery" is nondominated. If the battery is nondominated, add it to the list of Stage 2 batteries.

Stage n. To each battery of Stage $n - 1$, add a test and check whether the "new battery" is nondominated. If the battery is nondominated, add it to the list of Stage n batteries.

We can formulate this sequential process in the form of a dynamic program as follows. Let C_i ($i = 1, 2, \ldots, n$) represent the cost of test i and let P_i ($i = 1, 2, \ldots, n$) represent the performance of test i. The list of nondominated batteries at Stage 1 can be generated by treating the cost (C) as the state variable (see Section 2.4 for a description of the term "state") and finding the best performing test for each level of the state (cost). This can be given by solving the following equation over all feasible values for the state (cost):

Stage 1:

$$f_1^{k1(C)}(C) = \max_i (P_i) \quad \text{for all } i \text{ such that } C_i \leq C \quad (4.4.1)$$

We use the notation $k1(C)$ to represent the index (subscript) of the test that gives the best performance at the given level of cost, C, and $f_1^{k1(C)}$ to represent the performance of that particular test. If we solve Eq. (4.4.1) over all feasible cost values, we will end up with a list of all the nondominated tests.

The nondominated solutions for the remaining stages can be obtained by solving the following equation for each Stage m ($m = 2, \ldots, n$) over all feasible values of the state:

$$f_m^{km(C)}(C) = \max_i\{(P_i) \# [f_{m-1}^{km-1(C-C_i)}(C - C_i)]\},$$

$$i \notin km - 1(C - C_i), \quad C_i \le C \quad (4.4.2)$$

Note that in Eq. (4.4.2), $km(C)$ would be the set of indices of the tests in the battery that would solve Eq. (4.4.2) for a given level of the cost and $f_m^{km(C)}(C)$ would be the performance of that battery. We use the symbol $\#$ to represent the mathematical operation that is required to compute the performance of the battery at Stage m from (1) the performance of the battery at Stage $m - 1$ (i.e., $f_{m-1}^{km-1(C-C_i)}(C - C_i)$) and (2) the performance of the test that is added to that battery (i.e., P_i). Also note that we are assuming that the cost of the battery is given by the sum of the costs of the tests in the battery. In other words, if the cost of a particular battery is C, then we can compute the cost of this battery with the test T_i removed as the difference $C - C_i$. The restriction $i \notin km - 1(C - C_i)$ is used to ensure that the test that is added to the battery is not already present in the battery. When we solve Eq. (4.4.2) over all feasible values of the state (cost), we would like to obtain all of the nondominated batteries of size m or less, and consequently when we would reach the final stage (i.e., Stage n) we would have a list of all of the nondominated batteries. In order for this to happen, we must place some restrictions on the functional forms of the performance measures and cost functions.

Note that the formulation given in Eqs. (4.4.1) and (4.4.2) would require that we can (easily) compute the performance of the battery of m tests by using the value of the performance of $m - 1$ tests and the value of performance of the single additional test. This requirement is called *separability*, and it is required of all of the objective functions.

The commonly used cost functions can easily be shown to be separable. For example, if we compute the dollar cost of a battery as the sum of the costs of the individual tests, then we can see that the cost of a new battery that is formed by adding a new test to an old battery would be given by the cost of the old battery plus the cost of the new test—i.e., the additive cost function is separable. As a second example, suppose that the "time-cost" is given by the maximum of the times needed to complete each test. If we know the "time-cost" for the old battery and the time it takes to complete the new test, then the "time-cost" for the new battery would be given by the maximum of these two values.

In general, the measures of performance are not separable. For example, if we compute battery sensitivity using the positive majority rule and use this as our performance measure, then separability would require that we must be able to compute the sensitivity of a new battery of m tests as a function of (1) the sensitivity of the battery consisting of $m - 1$ tests and (2) the sensitivity of the test that is added to the battery. We can see that this would not be possible for two reasons: (1) we would need to know the sensitivities of each of the individual tests in the battery in order to compute the sensitivity of the new battery—the summary information given by $f_{m-1}^{km-1(C-C_i)}(C - C_i)$ along with the performance of the new test is not enough to compute the sensitivity of the new battery—and (2) we could not compute the sensitivity of a battery consisting of an even number of tests if we are using the majority rule.

In addition to separability, it can be shown that the objective functions must also be *serial-monotonic* with respect to preference, and preference over the attainable outcomes must be *nondominance bounded* in order for Eqs. (4.4.1) and (4.4.2) to generate all nondominated batteries (for proofs of this see Mitten, 1974; Villarreal and Karwan, 1982; Yu, 1985; and Pet-Edwards, 1986). We will not provide the details here, but we will indicate how these restrictions limit the form of the objective functions that can be used in the dynamic programming methodology.

Again we can show that the common measures of battery cost are serial-monotonic. Serial-monotonicity refers to the sequential nature of the problem and the direction of preference. In particular, if preference increases (decreases) with an increase in the value of the objective, then the objective value must also increase when you proceed to the next stage (i.e., when you add a new test to an existing battery) in order for an objective to be serially monotonic. We can easily see, for example, that additive cost would be serially monotonic—preference decreases as the cost increases, and cost increases as you proceed to the next stage.

Although most cost functions are both separable and serial-monotonic, this is not true of the performance functions. The restrictions of separability and serial-monotonicity on the objective functions limit the functional forms of the performance measures. We already noted that sensitivity based on positive majority rule is not separable. It turns out that of the performance measures described in Section 4.1, only the ones based on consensus rules would be separable under the assumption of statistical independence. For example, the sensitivity of a battery based on consensus would be simply given as the product of the sensitivities of the tests. Thus if we knew that the sensitivity of the $m-1$ test was β^+, then we would multiply β^+ by the sensitivity of the test that is added to the battery to get the sensitivity of the battery of m tests. A similar result can be stated for the specificity of the battery based on negative consensus. Recall, however, that the sensitivity and specificity measures computed based on the consensus rule should not be used to compare batteries composed of different numbers of tests (see Section 4.1). In particular, note that if we are comparing batteries with the same number of tests, then the battery with the highest sensitivity would be preferred; however, when one adds a test to a battery, the sensitivity (based on consensus) will always decrease—a battery consisting of more than one test would never be preferred over the single test! The sensitivity (as well as the specificity) based on consensus fails to be serially monotonic with preference. Here we have a performance measure where larger values are preferred, but the sequential process always produces smaller values for the measure. The direction of preference and the direction produced by the sequential process are in conflict.

We noted above that only the performance measures that are based on consensus rule would be separable. We found, however, that the sensitivity and specificity of a battery fail to be serially monotonic and consequently cannot be used in the dynamic programming method for battery construction. The only other measures based on consensus rule described in Section 4.1 are the predictivity-based measures. It is not intuitively obvious, however, that these measures are separable or serially monotonic. In order to see that the measures are separable, suppose that we have m tests, with test i having a sensitivity of α_i^+ and a specificity of α_i^-. Suppose also that the tests are conditionally independent. We can compute the predictivity of the battery for the case where all of the tests gave positive results as follows:

$$\Pr(P|+_1, +_2, \ldots, +_m)$$

$$= \frac{1}{1 + \dfrac{\Pr(NP)(1-\alpha_1^-)(1-\alpha_2^-)\cdots(1-\alpha_m^-)}{\Pr(P)(\alpha_1^+)(\alpha_2^+)\cdots(\alpha_m^+)}} \quad (4.4.3)$$

Similarly, the predictivity for all negative results would be given by:

$$\Pr(NP|-_1, -_2, \ldots, -_m)$$

$$= \frac{1}{1 + \dfrac{\Pr(P)(1-\alpha_1^+)(1-\alpha_2^+)\cdots(1-\alpha_m^+)}{\Pr(NP)(\alpha_1^-)(\alpha_2^-)\cdots(\alpha_m^-)}} \quad (4.4.4)$$

Battery Selection

Suppose that we add a new test to the battery. Let us see what happens to the predictivity based on positive consensus. Note that Eq. (4.4.3) would be altered as follows:

$$\Pr(P|+_1, +_2, \ldots, +_m, +_{m+1})$$

$$= \frac{1}{1 + \dfrac{\Pr(NP)(1 - \alpha_1^-)(1 - \alpha_2^-) \cdots (1 - \alpha_m^-)(1 - \alpha_{m+1}^-)}{\Pr(P)(\alpha_1^+)(\alpha_2^+) \cdots (\alpha_m^+)(\alpha_{m+1}^+)}} \quad (4.4.5)$$

(A similar equation can be given for $m + 1$ negative results.) If we denote the predictivity based on m tests given by Eq. (4.4.3) by P_m^+ and the predictivity based on $m + 1$ tests by P_{m+1}^+, then the relationship between Eqs. (4.4.3) and (4.4.5) can be written as follows:

$$P_{m+1}^+ = \frac{1}{1 + \dfrac{(1 - P_m^+)(1 - \alpha_{m+1}^-)}{(P_m^+)(\alpha_{m+1}^+)}} \quad (4.4.6)$$

Similarly, if we let P_m^- and P_{m+1}^- represent the predictivities based on m and $m + 1$ negative test results, respectively, then the relationship can be expressed as follows:

$$P_{m+1}^- = \frac{1}{1 + \dfrac{(1 - P_m^-)(1 - \alpha_{m+1}^+)}{(P_m^-)(\alpha_{m+1}^-)}} \quad (4.47)$$

From the expressions given in Eqs. (4.4.6) and (4.4.7), we can see that the predictivity measures based on the consensus rule are separable—i.e., we can compute the predictivities of the new battery (i.e., P_{m+1}^+ and P_{m+1}^-) based simply on the knowledge of the predictivity of the old battery (i.e., P_m^+ and P_m^-), along with knowledge about the performance of the test that is added to the current battery (i.e., α_{m+1}^+ and α_{m+1}^-).

In order to use the predictivities in the dynamic programming formulation, we need to check whether they are serial-monotonic. In particular, since batteries with larger values for the predictivities would be preferred, in order for the predictivities to be serial monotonic we require that $P_{m+1} \geq P_m$ (i.e., the predictivity must improve when a test is added to the battery). If we compare Eqs. (4.4.3) and (4.4.5), we can see that this would require that the following relationships must hold between the sensitivities and specificities of the test that is added to the battery:

$$\frac{(1 - \alpha_{m+1}^-)}{\alpha_{m+1}^+} \leq 1.0 \quad \text{and} \quad \frac{(1 - \alpha_{m+1}^+)}{\alpha_{m+1}^-} \leq 1.0 \quad (4.4.8)$$

or equivalently

$$1 - \alpha^-_{m+1} \leq \alpha^+_{m+1} \tag{4.4.9}$$

Thus, in the case in which we are dealing with conditionally independent test results, the predictivity measures computed using the consensus rule are separable and are serially monotonic as long as the sensitivities and specificities of the tests satisfy Eq. (4.4.8). Note that a sufficient condition for Eq. (4.4.8) would be that the sensitivities and the specificities of the tests exceed 0.5—which is not an unreasonable restriction.

For the remainder of this section, we will assume that we are using the battery performance measures defined in Eqs. (4.4.3) and (4.4.4), that the sensitivities and specificities satisfy Eq. (4.4.9), and that the individual test performances are represented by two ratios given in Eq. (4.4.8). Thus by solving Eqs. (4.4.1) and (4.4.2) from Stage 1 to Stage n (the number of tests) we are guaranteed to construct all of the nondominated batteries.

In the dynamic programming formulation given in Eqs. (4.4.1) and (4.4.2), we noted that the equation at each stage should be solved over all feasible values of the cost function. We can determine the feasible ranges of the costs at each stage by looking at the costs of the individual tests. In particular, suppose we order the costs of the individual tests from minimum to maximum and define $C_m^{\min}$ to be the sum of the m lowest cost tests and $C_m^{\max}$ to be the sum of the m highest cost tests. Now consider the computations at Stage m (where we are considering batteries of size m or less). Note that all of the nondominated batteries constructed at Stage $m - 1$ that cost less than $C_m^{\min}$ must still be nondominated at Stage m. This is because the battery of tests associated with $C_m^{\min}$ would be the least costly battery consisting of m tests regardless of its performance, which implies that all of the batteries that cost less than $C_m^{\min}$ would consist of less than m tests, and all of the nondominated batteries consisting of less than m tests were constructed at Stage $m - 1$. We can also see that the maximum cost at Stage m would be given by $C_m^{\max}$. Thus at Stage m, we need only compute new batteries for state values (costs) between $C_m^{\min}$ and $C_m^{\max}$.

In a similar manner we can also compute the minimum and maximum ranges for the performance measures. For example, suppose we are using the predictivity measure based on positive test results [see Eq. (4.4.3)]. Let the performance of test i be represented by the ratio of one minus the specificity of the test to the sensitivity of the test [refer to Eq. (4.4.8)]. We will call this the "positive performance ratio" for test i. Note that the maximum that P_m^+ can achieve would be given by the m tests that have the smallest "positive performance ratios." Denote this by $P_m^{+\max}$. The minimum that P_m^+ can be would be given by the m tests that have the largest "positive

Battery Selection

performance ratios." Denote this by $P_m^{+\min}$. Thus at each stage of the battery construction, we have already constructed the nondominated solutions corresponding to performances less than $P_m^{+\min}$ at Stage $m - 1$, and we do not have to consider any values of the performance greater than $P_m^{+\max}$. We can define the minimum and maximum ranges for P_m^- in an analogous fashion, where the "negative performance ratio" for each test would be given by the ratio of one minus the sensitivity of the test to its specificity [refer to the second expression given in Eq. (4.4.8)].

Using the preceding bounds on the state (cost) and the objective (performance), we can now rewrite the dynamic programming formulation as follows:

Stage 1:

$$f_1^{k1(C)}(C)$$

$$= \max_i(P_i) \quad \text{for all } i \text{ such that } C_i \leq C \leq C_1^{\max} \quad (4.4.10)$$

Stage m:

$$f_m^{km(C)}(C)$$

$$= \max_i\{(P_i) \# (f_{m-1}^{km-1(C-C_i)}(C - C_i))\}, \quad i \notin km - 1(C - C_i),$$

$$C_i \leq C \quad \text{where } C_m^{\min} \leq C \leq C_m^{\max} \quad \text{and} \quad P_m^{\min} \leq f_m^{km(C)}(C) \leq P_m^{\max}$$

where either P_i now represents the positive performance ratio of test i with the functional relationship "$\#$" given by Eq. (4.4.6) or P_i represents the negative performance ratio of test i with the functional relationship "$\#$" given by Eq. (4.4.7).

Suppose we would like to find a battery that would maximize the predictivity of a consensus of positive results and would minimize the cost. By applying Eqs. (4.4.10) and (4.4.11) we can construct all of the nondominated batteries. The computations would proceed as follows if we have a total of n tests to consider.

DYNAMIC PROGRAMMING FOR CONSTRUCTION OF NONDOMINATED BATTERIES

Step 1. For $i = 1$ to n compute $C_i^{\min}$ as the sum of the i lowest cost tests and $C_i^{\max}$ as the sum of the i highest cost tests.

Step 2. Compute the positive performance ratio of each test (i.e., the ratio of one minus the specificity to the sensitivity. Then for $i = 1$ to n compute P_i^{max} as the value of Eq. (4.4.3) using the i tests with the lowest positive performance ratios, and compute P_i^{min} as the value of Eq. (4.4.3) using the i tests with the largest performance ratios.

Step 3. Solve the Stage 1 problem using Eq. (4.4.10). Start with the lowest value for the state (i.e., a cost of zero) and solve the equation for successively larger values of the cost.

Step 4. For Stages 2 through n solve Eq. (4.4.11). For example, at Stage m, first list all of the nondominated batteries from the previous stage that correspond to $C < C_m^{min}$. Begin the computations with a value for the state corresponding to $C \geq C_m^{min}$ and solve Eq. (4.4.11) for successively larger values of the cost, thus generating the list of nondominated batteries of size m or less. When the solution to Eq. (4.4.11) is a battery with performance equal to P_m^{max} then we proceed to the next stage.

In order to implement the preceding algorithm we would discretize the state values. That is, we would solve Eqs. (4.4.10) and (4.4.11) over discrete intervals of the cost. If, for example, the costs were given in terms of hundreds of dollars, we would solve the equations for every hundred dollar increment of the cost. If the costs were given in terms of thousands of dollars, we would solve the equations for every thousand dollar increment of cost.

EXAMPLE 4.4.1

As a simple illustration of this procedure for constructing the nondominated batteries, consider the five tests given in Table 4.4.1. The last two columns of the table show the performance ratios for each test. The constraints on the cost and performance indices for each stage of the dynamic program are given in Table 4.4.2.

TABLE 4.4.1. Costs, Sensitivities, Specificities, and Performance Ratios of the Tests Used in Example 4.4.1

i	Tests	Costs ($)	Sensitivities	Specificities	P_i^+	P_i^-
1	T_1	500	0.8	0.6	0.67	0.75
2	T_2	600	0.8	0.55	0.64	0.74
3	T_3	1000	0.6	0.88	0.83	0.69
4	T_4	100	0.6	0.52	0.56	0.56
5	T_5	200	0.55	0.505	0.53	0.53

Battery Selection

TABLE 4.4.2. Constraints Used in Example 4.4.1

Stage	C^{min}	P^{min+}	P^{min-}
1	100	0.833	0.752
2	300	0.909	0.894
3	800	0.947	0.949
4	1400	0.957	0.960
5	2400	0.961	0.965

TABLE 4.4.3. Results for Stage 1: Performance Based on Prediction of P^+

Cost ($)	Performance	Battery
100	0.556	4
500	0.667	1
1000	0.833	3

Suppose we are interested in finding the "best" battery that would minimize cost and maximize the positive performance ratio. Note that the costs are given in terms of hundreds of dollars; thus at Stage 1, we will solve Eq. (4.4.10) for every hundred dollar increment of cost. The nondominated tests for Stage 1 (i.e., batteries of size one) are given in Table 4.4.3 and also displayed in Fig. 4.4.1. At Stage 2, where batteries of size two or

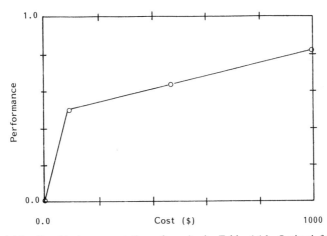

FIGURE 4.4.1. Graphical representation of results in Table 4.4.3. Optimal frontier for Stage 1.

TABLE 4.4.4. Results for Stage 2: Performance Based on Prediction of P^+

Cost ($)	Performance	Battery
100	0.556	4
300	0.581	4, 5
500	0.667	1
600	0.714	1, 4
1000	0.833	3
1100	0.862	3, 4
1500	0.909	3, 1

smaller are considered, the nondominated batteries are computed using Eq. (4.4.11) with $m = 2$ over every feasible $100 increment of cost. These results are summarized in Table 4.4.4 and displayed in Fig. 4.4.2. Notice that in Table 4.4.4 the Stage 2 solutions are identical to the Stage 1 solutions for costs less than $300 ($300 = $C_2^{\min}$) and that the Stage 2 solution stops after reaching $P_2^{\max}$.

The following is a sample calculation for Stage 2. Suppose we wish to solve Eq. (4.4.11) for $C = 1100$ (i.e., cost = $1100):

$$f_2^{k2(1100)}(1100) = \max\{(P_i^+) \# [f_1^{k1(1100-C_i)}(1100 - C_i)]\}$$

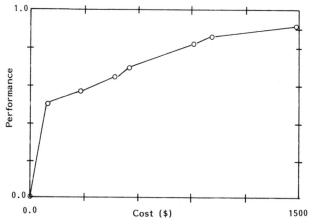

FIGURE 4.4.2. Graphical representation of results in Table 4.4.4. Optimal frontier for Stage 2.

Battery Selection

We can expand the right-hand side of the preceding equation over every possible test, namely,

$$P_0^+ \# f_1(1100) = 0.83$$

$$P_1^+ \# f_1(600) = \text{infeasible} \quad (T_1 \text{ is already used})$$

$$P_2^+ \# f_1(500) = 0.78$$

$$P_3^+ \# f_1(100) = 0.86$$

$$P_4^+ \# f_1(1000) = 0.86$$

$$P_5^+ \# f_1(900) = 0.69$$

(We have used the notation P_0^+ to denote that no test is added.) Note that the maximum value is given by $P_3^+ \# f_1(100) = 0.86$ and $P_4^+ \# f_1(1000) = 0.86$, where the battery consists of T_3 and T_4. Thus $k_2(1100) = \{3, 4\}$. The best performance that we can achieve for a battery costing $1100 or less would be $P^+ = 0.86$ using T_3 and T_4.

This same procedure is carried out over all values of the state and over the remaining stages. These results are summarized in Tables 4.4.5–4.4.7 and Figs. 4.4.3–4.4.5. Notice that after the completion of the last stage, we obtain a list of all of the nondominated batteries, conveniently ordered in terms of increasing cost and increasing performance. The decision maker can now choose his/her most preferred battery from among these by using the procedures of Section 4.2.

TABLE 4.4.5. Results for Stage 3: Performance Based on Prediction of P^+

Cost ($)	Performance	Battery
100	0.556	4
300	0.581	4, 5
500	0.667	1
600	0.714	1, 4
800	0.735	1, 4, 5
1000	0.833	3
1100	0.862	3, 4
1300	0.874	3, 4, 5
1500	0.909	3, 1
1600	0.926	3, 1, 4
2100	0.947	3, 1, 2

TABLE 4.4.6. Results for Stage 4: Performance Based on Prediction of P^+

Cost ($)	Performance	Battery
100	0.556	4
300	0.581	4, 5
500	0.667	1
600	0.714	1, 4
800	0.735	1, 4, 5
1000	0.833	3
1100	0.862	3, 4
1300	0.874	3, 4, 5
1500	0.909	3, 1
1600	0.926	3, 1, 4
1800	0.933	3, 4, 1, 5
2100	0.947	3, 1, 2
2200	0.957	3, 1, 2, 4

The dynamic programming method for battery construction has been presented and illustrated for the case in which we are interested in two objectives. The method, however, can be easily extended to more than two objectives by using multistate dynamic programming.

TABLE 4.4.7. Results for Stage 5: Performance Based on Prediction of P^+

Cost ($)	Performance	Battery
100	0.556	4
300	0.581	4, 5
500	0.667	1
600	0.714	1, 4
800	0.735	1, 4, 5
1000	0.833	3
1100	0.862	3, 4
1300	0.874	3, 4, 5
1500	0.909	3, 1
1600	0.926	3, 1, 4
1800	0.933	3, 4, 1, 5
2100	0.947	3, 1, 2
2200	0.957	3, 1, 2, 4
2400	0.961	1, 2, 3, 4, 5

Battery Selection

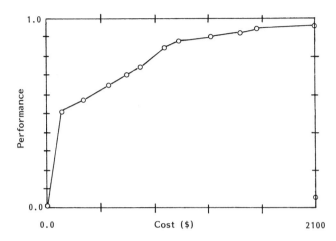

FIGURE 4.4.3. Graphical representation of results in Table 4.4.5. Optimal frontier for Stage 3.

Suppose we want to apply the tests sequentially, and we want to find a battery that would have minimum cost, maximum performance (given by one of the two predictivity measures computed by consensus rule), and minimum testing time. Since the tests are to be applied sequentially, the testing time for a battery of tests would be given by the sum of the testing times of the individual tests. If we let t_i be the time it takes to complete

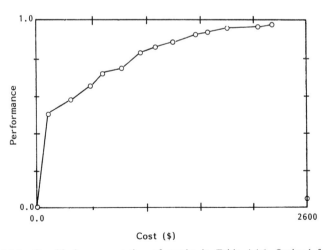

FIGURE 4.4.4. Graphical representation of results in Table 4.4.6. Optimal frontier for Stage 4.

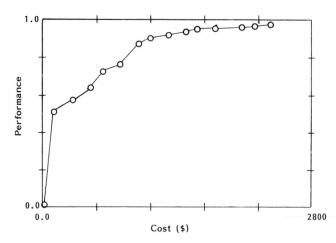

FIGURE 4.4.5. Graphical representation of results in Table 4.4.7. Optimal frontier for Stage 5.

test i, let t represent time as a state variable, and use the same notation as given earlier, we can give the dynamic programming formulation as follows:

Stage 1:

$$f_1^{k1(C,t)}(C, t) = \max_i(P_i) \qquad \text{for all } i \text{ such that } C_i \leq C \text{ and } t_i \leq t$$

Stage m:

$$f_m^{km(C,t)}(C, t) = \max_i\{(P_i) \# [f_{m-1}^{km-1(C-C_i, t-t_i)}(C - C_i, t - t_i)]\},$$

$$i \notin km - 1(C - C_i), \qquad C_i \leq C \quad \text{and} \quad t_i \leq t$$

If at each stage we solve the preceding equations over all values of the two states, C and t, we will generate a list of all of the nondominated batteries after completing Stage n. We can generalize this process to any number of objectives by adding an additional state for each additional objective, but it should be noted that consideration of too many objectives at one time would make the procedure computationally intensive and consequently would lessen its computational advantage.

We will end this section with a discussion on the computational efficiency of the dynamic programming-based method. We have seen that this battery construction method will produce only batteries that are nondominated. This can result in a considerable savings in computation over the complete enumeration of all of the batteries, where we would (1)

Battery Selection 163

construct every battery, (2) compute the performance of every battery, and (3) compare every pair of the batteries in order to reduce the number to only those that are nondominated. In order to obtain an idea of the potential computational savings, let us first consider how many computations would be required by the complete enumeration process and then compare this with the number of computations required in the dynamic programming-based process. Complete enumeration would require the computation of the performances of $2^n - 1$ batteries and would require $(2^n - 1)(2^n)/2$ comparisons to find the nondominated solutions. Thus an estimate for the total number of computations for the complete enumeration process would be

$$N_{CE} = (2^n - 1) + (2^n - 1)(2^n)/2$$

To get an idea of the approximate number of computations required by the dynamic programming-based method, we note that for n tests, the method has n stages; and for each stage, we apply the computations for each value of the state. Thus the number of computations would be given by the product of (1) the number of states per stage summed over the number of stages and (2) the number of computations per state. Recall that we discretized the state in order to do the computations. If ΔC represents the cost increment (e.g., 100 if we are dealing with hundreds of dollars, or 1000 if we are dealing with thousands of dollars), then the number of state values requiring computation in Stage m would be given by $C_m^{max}/\Delta C$. (Note that this is actually an overestimate since we only need to do computations starting at C_m^{min}.) The total number of states requiring computation would then be given by the sum of these ratios over all stages. Note that the value of $C_m^{max}/\Delta C$ will increase with m and is dependent on the specific cost of each test. To simplify the computation a bit, let C_{av} be the average cost of a test. The approximate cost of m tests would then be given by mC_{av}. Since the nth stage (where all tests are considered) requires no computation, we can estimate the number of states requiring computation as the sum of $mC_{av}/\Delta C$ from $m = 1$ to $n - 1$. This reduces to $(n)(n-1)C_{av}/2\Delta C$. For each value of the state, we compute approximately n batteries. Thus we can estimate the number of computations required by the dynamic programming-based method as

$$N_{DP} = \frac{(n^2)(n-1)C_{av}}{2\Delta C}$$

Table 4.4.8 illustrates the approximate number of computations required by the dynamic programming-based method as a function of the average test cost and the number of tests (we assume that $\Delta C = 100$ for all

TABLE 4.4.8. Comparison on the Number of Computations Required by the Dynamic Programming-Based Method versus Complete Enumeration

	D.P. procedure: C				Direct procedure:
n	500	1000	1500	2000	any $\bar{C}$
5	250	500	750	1,000	527
10	2,250	4,500	6,750	9,000	524,799
15	7,875	15,750	23,625	31,500	536,857,297
20	19,000	38,000	57,000	76,000	5.5×10^{11}
25	37,500	75,000	112,500	150,000	5.63×10^{14}
30	65,250	130,500	195,750	261,000	5.76×10^{17}

of the computations), and compare these numbers with the number of computations required by the complete enumeration process. Note that as the number of tests increases, the number of computations required by complete enumeration increases drastically (e.g., doubling the number of tests requires over one million times the number of computations). However, the computational effort required by dynamic programming is much less, especially when the number of tests is large.

REFERENCES

Bellman, R. E., and Dreyfus, S. E., 1962, *Applied Dynamic Programming*, Princeton University Press, Princeton, New Jersey.
Heinze, J. E., and Poulsen, N. K., 1983, "The optimal design of batteries of short-term tests for detecting carcinogens," *Mutation Res.*, **117**:259-269.
McCann, J., Choi, E., Yamasaki, E., and Ames, B. N., 1975, "Detection of carcinogens as mutagens in the Salmonella/microsome test: Assay of 300 chemicals," *Proc. Natl. Acad. Sci. (U.S.A.)*, **72**:5135-5139.
Mitten, L. G., 1974, "Preference order dynamic programming," *Management Sci.*, **21**:43-46.
Pet-Edwards, J., 1986, "Selection and interpretation of conditionally dependent tests for binary prediction: A Bayesian approach," Ph.D. dissertation, Case Western Reserve University, Cleveland, Ohio.
Pet-Edwards, J., Chankong, V., Rosenkranz, H. S., and Haimes, Y. Y., 1985, "Application of the carcinogenicity prediction and battery selection (CPBS) method to the Gene-tox data base," *Mutation Res.*, **153**:187-200.
Tauxe, G. W., Inman, R. R., and Mades, D. M., 1979, "Multiobjective dynamic programming: A classic problem redressed," *Water Resource Res.*, **15**:1398-1402.
Villarreal, B., and Karwan, M. H., 1982, "Multicriteria dynamic programming with an application to the integer case," *J. Optim. Theory Appl*, **38**:43-69.
Yu, P.-L., 1985, *Multiple-Criteria Decision Making*, Plenum Press, New York.

Chapter 5

Risk Assessment Using Test Results

Consider the situation in which we would like to determine whether some object has a certain property. For example, the object might be a chemical and the unknown property might be the potential carcinogenicity of the chemical. Suppose, further, that we have a set of tests that we can use to help us determine whether the property is present in the object. In Chapter 3 we discussed several analyses for computing the performances of the tests and the interdependencies between the pairs of tests, and in Chapter 4 we discussed how one can select the "best" battery of tests to use for a particular application. In this chapter, it is assumed that we have chosen a battery to use, we have applied this battery on the object, and now we must interpret the results of this battery.

For our purposes, this battery prediction problem can be formally defined as follows. Suppose we are given the following information:

1. The results of a battery of k tests;
2. Estimates for the sensitivities and specificities of the k tests; and
3. Estimates for the conditional dependencies between each pair of tests.

The goal of battery prediction is to interpret the combined k test results in order that we can decide whether the object has or does not have the property of interest. The interpretation is in the form of a probability which indicates the potential presence of the property in the tested object.

We will consider two cases. In Section 5.1 we will discuss how to make predictions when the tests are conditionally independent (or as an approximation when each pair of tests is conditionally independent). We develop and demonstrate both "batch" predictions (i.e., predictions based on using all of the tests at the same time) and sequential predictions using Bayes' theorem. In Section 5.2, we will discuss how the Bayesian predictions can be made when tests are conditionally dependent. We will demonstrate how the pairwise K_P measures can be used to approximate these predictions.

5.1. MAKING PREDICTIONS WHEN TESTS ARE CONDITIONALLY INDEPENDENT

Suppose that we have k test results, $r_1, r_2, \ldots, r_k$ and that from preliminary analysis we have found that the tests are conditionally independent (see Chapter 3). Let $\alpha_1^+, \alpha_2^+, \ldots, \alpha_k^+$ be the sensitivities of the tests, $\alpha_1^-, \alpha_2^-, \ldots, \alpha_k^-$ be the specificities of the tests. Recall that Bayes' formula can be given as function of the sensitivities and specificities of the tests when the tests are independent (see Chapter 4). In particular, Bayes' formula can be given as

$$\Pr(P|r_1, r_2, \ldots, r_k) = \frac{1}{1 + \dfrac{\Pr(NP)\Pr(r_1, r_2, \ldots, r_k|NP)}{\Pr(P)\Pr(r_1, r_2, \ldots, r_k|P)}} \quad (5.1.1)$$

When the tests are conditionally independent, then the joint probabilities can be given as the product of the marginal probabilities, namely

$$\Pr(P|r_1, r_2, \ldots, r_k) = \frac{1}{1 + \dfrac{\Pr(NP)\Pr(r_1|NP) \cdots \Pr(r_k|NP)}{\Pr(P)\Pr(r_1,|P) \cdots \Pr(r_k|P)}} \quad (5.1.2)$$

Note that in Eq. (5.1.2), when r_i is positive, then $\Pr(r_i|P)$ would be the sensitivity of test i and $\Pr(r_i|NP)$ would be one minus the specificity of test i. Similarly, if r_i was negative, then $\Pr(r_i|P)$ would be one minus the sensitivity of test i and $\Pr(r_i|NP)$ would be the specificity of test i. Equation (5.1.2) gives us the *batch formulation* for the Bayesian prediction. In other words, it allows us to make the Bayesian prediction based on the simultaneous (batch) interpretation of all k tests.

Risk Assessment Using Test Results

As an illustration, suppose we have results from six tests with sensitivities, specificities, and results given in Table 5.1.1. The prediction of the property would be given by Eq. (5.1.2) as follows:

$$\Pr(P \mid +_1, +_2, +_3, -_4, +_5, +_6)$$

$$= \frac{1}{1 + \dfrac{\Pr(NP)(0.20)(0.14)(0.50)(0.80)(0.05)(0.05)}{\Pr(P)(0.61)(0.61)(0.65)(0.20)(0.56)(0.76)}}$$

$$= \frac{1}{1 + \dfrac{\Pr(NP)(0.000028)}{\Pr(P)(0.02059)}}$$

If we have an estimate for the prior (before testing) probability that the object has the property, then we can use the preceding expression to compute the probability that the object has the property given the results of the six tests (i.e., the posterior probability). For example, if the prior probability, $\Pr(P)$, is 0.2, then the posterior probability would be 0.995; and if the prior probability is 0.5, then the posterior probability would be 0.9986. Figure 5.5.1 gives the range of values of the posterior probability as a function of the prior probability.

The predictions may also be given sequentially. It is easy to show that the following equation is equivalent to Eq. (5.1.2):

$$\Pr(P \mid r_1, r_2, \ldots, r_k)$$

$$= \frac{\Pr(P \mid r_1, \ldots, r_{k-1})\Pr(r_k \mid P)}{\Pr(P \mid r_1, \ldots, r_{k-1})\Pr(r_k \mid P) + \Pr(NP \mid r_1, \ldots, r_{k-1})\Pr(r_k \mid NP)} \quad (5.1.3)$$

TABLE 5.1.1. Sensitivities and Specificities of Five Tests

Test	Sensitivity	Specificity	Result
T_1	0.61	0.80	+
T_2	0.61	0.86	+
T_3	0.65	0.50	+
T_4	0.80	0.80	−
T_5	0.56	0.95	+
T_6	0.76	0.95	+

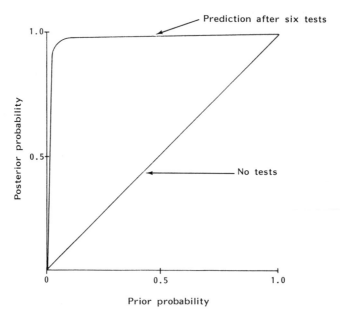

FIGURE 5.1.1. Range of posterior batch predictions as a function of the prior probability. (Data as given in Table 5.1.1.)

We can rewrite Eq. (5.1.3) in terms of the sensitivities and specificities of the tests. In particular, let θ_i^+ be the estimate for the posterior probability of P after using i of the tests (i.e., $\theta_i^+ = \Pr(P|r_1, r_2, \ldots, r_i)$); and let θ_0^+ represent the prior probability of P [i.e., $\Pr(P)$ using zero tests]. Suppose that our first test gives a positive result on the object. We can write Eq. (5.1.3) as follows:

$$\Pr(P|+_1) = \theta_1^+ = \frac{\theta_0^+ \alpha_1^+}{(1 - \alpha_1^-) + (\alpha_1^+ + \alpha_1^- - 1)\theta_0^+} \tag{5.1.4}$$

Similarly, if T_1 gave a negative result, Eq. (5.1.3) would be

$$\Pr(P|-_1) = \theta_1^+ = \frac{\theta_0^+ (1 - \alpha_1^+)}{\alpha_1^- - (\alpha_1^+ + \alpha_1^- - 1)\theta_0^+} \tag{5.1.5}$$

We can extend Eqs. (5.1.4) and (5.1.5) to i tests as follows. Suppose we know the prediction based on $i - 1$ tests (i.e., θ_{i-1}^+). If the ith test is positive, we would compute the Bayesian prediction as follows:

$$\theta_i^+ = \frac{\theta_{i-1}^+ \alpha_i^+}{(1 - \alpha_i^-) + (\alpha_i^+ + \alpha_i^- - 1)\theta_{i-1}^+} \tag{5.1.6}$$

Risk Assessment Using Test Results

and if the *i*th test was negative, then we would compute the Bayesian prediction as

$$\theta_i^+ = \frac{\theta_{i-1}^+ (1 - \alpha_i^+)}{\alpha_i^- - (\alpha_i^+ + \alpha_i^- - 1)\theta_{i-1}^+} \quad (5.1.7)$$

Equations (5.1.6) and (5.1.7) give us the *sequential formulation* for the Bayesian prediction. In other words, we can compute the series of Bayesian predictions given by the sequence of tests.

As an illustration of how the sequential formulas given in Eqs. (5.1.6) and (5.1.7) can be used, consider the example presented earlier (refer to the data given in Table 5.1.1). Suppose the prior probability of P is 0.5. The first test gave a positive result. Thus we use Eq. (5.1.6) with $i = 1$, $\alpha_1^+ = 0.61$, $\alpha_1^- = 0.80$, and $\theta_0^+ = 0.5$. This gives us the following:

$$\theta_1^+ = (0.5)(0.61)/[(1 - 0.80) + (0.61 + 0.80 - 1)(0.50)]$$

$$= 0.753$$

The second test gave a positive result. Thus we use Eq. (5.1.6) with $i = 2$, $\alpha_2^+ = 0.61$, $\alpha_2^- = 0.86$, and $\theta_1^+ = 0.753$. This gives us a value for θ_2^+ of 0.93. Our third test was also positive for the property and gives us a value of $\theta_3^+ = 0.945$. (Note that the values of the sensitivity and specificity of the third test are rather low. Thus the result of the third test did not have very much effect on the prediction.) The fourth test was negative for the property. Thus we use Eq. (5.1.7) with $i = 4$, $\alpha_4^+ = 0.80$, $\alpha_4^- = 0.80$, and $\theta_3^+ = 0.945$. The computation would proceed as follows:

$$\theta_4^+ = (0.945)(0.20)/[0.80 - (0.80 + 0.80 - 1)(0.945)]$$

$$= 0.811$$

Continuing in this fashion, we get $\theta_5^+ = 0.980$ and, finally, $\theta_6^+ = 0.9986$ (the same result as the batch formulation). We can also give the predictions based on this sequence of test results as a function of the prior probability. These results are displayed in Fig. 5.1.2.

The preceding batch and sequential formulations for Bayesian predictions can be used when the tests are conditionally independent. We also indicated that as long as each pair of tests is conditionally independent and the battery is not too large (say less than 10 tests), then the tests can (as an approximation) be treated as being conditionally independent (see Section 4.1). Recall, however, that when the tests are known to be conditionally dependent, then the use of the preceding equations could result in

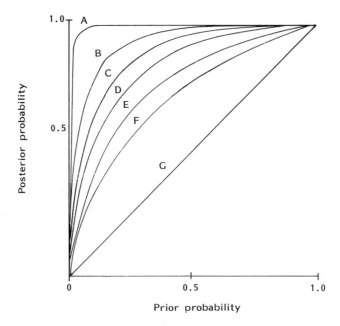

FIGURE 5.1.2. Range of posterior sequential predictions as a function of the prior probability. (Data as given in Table 5.1.1.) A, after six tests; B, after five tests; C, after three tests; D, after two tests; E, after four tests; F, after one test; and G, with no tests.

considerable errors in the predictions (refer to Section 3.4). The following section describes how the Bayesian predictions can be adjusted for the dependencies between the tests.

5.2. MAKING PREDICTIONS WHEN TESTS ARE CONDITIONALLY DEPENDENT

Suppose that preliminary analysis of the past data on the tests has indicated that some of the tests are conditionally dependent. Recall that we can compute the conditional dependence between two tests as

$$K_P(r_1, r_2) = \frac{\Pr(r_1, r_2 | P)}{\Pr(r_1 | P) \Pr(r_2 | P)} \qquad (5.2.1)$$

Risk Assessment Using Test Results

and

$$K_{NP}(r_1, r_2) = \frac{\Pr(r_1, r_2 | NP)}{\Pr(r_1 | NP)\Pr(r_2 | NP)} \quad (5.2.2)$$

And we can compute the dependence between k tests as

$$K_P(r_1, \ldots, r_k) = \frac{\Pr(r_1, \ldots, r_k | P)}{\Pr(r_1 | P) \cdots \Pr(r_k | P)} \quad (5.2.3)$$

and

$$K_{NP}(r_1, \ldots, r_k) = \frac{\Pr(r_1, \ldots, r_k | NP)}{\Pr(r_1 | NP) \cdots \Pr(r_k | NP)} \quad (5.2.4)$$

Using these measures, we can write the joint probabilities given in Eq. (5.1.1) as follows:

$$\Pr(r_1, r_2, \ldots, r_k | NP) = K_{NP}(r_1, \ldots, r_k)\Pr(r_1 | NP) \cdots \Pr(r_k | NP)$$

$$\Pr(r_1, r_2, \ldots, r_k | P) = K_P(r_1, \ldots, r_k)\Pr(r_1 | P) \cdots \Pr(r_k | P)$$

Thus the batch formulation can be given as

$$\Pr(P | r_1, r_2, \ldots, r_k)$$

$$= \frac{1}{1 + \dfrac{\Pr(NP) K_{NP}(r_1, \ldots, r_k)\Pr(r_1 | NP) \cdots \Pr(r_k | NP)}{\Pr(P) K_P(r_1, \ldots, r_k)\Pr(r_1, | P) \cdots \Pr(r_k | P)}} \quad (5.2.5)$$

Note that Eq. (5.2.5) requires values for the K_P measures given in Eqs. (5.2.3) and (5.2.4). This would in turn require that we have adequate data on the joint responses of the tests on a large number of objects with known properties. This is often not available. Suppose, however, that we do have information on the conditional dependencies between pairs of tests [i.e., information of the type given in Eqs. (5.2.1) and (5.2.2) for each pair of tests]. An approximation for the K_P measures given in Eqs. (5.2.3) and (5.2.4) based on the pairwise K_P measures was discussed earlier (see Pet-Edwards, 1986; also refer to Section 4.1). This approximation was given by the product of the $(k - 1)$ pairwise K_P measures (for positive results) that deviate the most from the value one. Thus we can replace $K_P(r_1, \ldots, r_k)$ by the product of $(k - 1)$ values of $K_P(r_i, r_j)$ and $K_{NP}(r_1, \ldots, r_k)$ by the product of $(k - 1)$ values of $K_{NP}(r_{i,j})$.

TABLE 5.2.1. Hypothetical Data on Four Dependent Tests

Tests	Sensitivity	Specificity	Result
T_1	0.80	0.80	+
T_2	0.70	0.85	+
T_3	0.95	0.65	+

TABLE 5.2.2. Pairwise KP Measures for Hypothetical Tests in Table 5.2.1

Test pair	$K_P(++)$	$K_{NP}(++)$
T_1, T_2	1.25	1.1
T_1, T_3	0.50	1.0
T_2, T_3	1.00	1.1

To illustrate the preceding equations, consider the hypothetical data on three tests given in Table 5.2.1. Suppose, further, that preliminary analysis gives us the K_P measure given in Table 5.2.2. Then using the preceding estimate for the joint K_P measures based on the pairwise K_P measures, the prediction based on Eq. (5.1.5) would be given by

$$\Pr(P|+_1, +_2, +_3) = \frac{1}{1 + \dfrac{\Pr(NP)(1.1)(1.1)(0.2)(0.15)(0.35)}{\Pr(P)(1.25)(0.5)(0.8)(0.7)(0.95)}}$$

In this chapter, we have discussed and illustrated how a battery of test results (either statistically dependent or independent) can be interpreted using Bayes' formula. The computations involve knowledge about the past performances of the tests (i.e., the sensitivities and specificities), the conditional dependencies among the tests, and a prior probability that the property is present in the tested object. The result of this analysis would be a probability that the property is present in the tested object [perhaps given as a function of the prior (before testing) probability that the property is present in the object] based on the results of the tests on the given object. It is now up to the decision maker to decide whether the resulting probability is high or low enough to base his decision on, or whether more testing may be needed.

We can now see how preliminary analysis, battery selection, and Bayesian prediction each form an integral part of the CPBS approach (the reader should refer to Fig. 1.1.1 in Chapter 1). The objective of this book has been to provide the reader with a theoretically sound approach to selecting and interpreting the results of multiple tests. The following chapter is used to further illustrate many of the concepts from Chapter 4 (Battery Selection) and Chapter 5 (Bayesian Prediction). It is an application of the CPBS approach to the cancer hazard identification problem.

REFERENCE

Pet-Edwards, J., 1986, "Selection and interpretation of conditionally dependent tests for binary predictions: A Bayesian approach," Ph.D. dissertation, Case Western Reserve University, Cleveland, Ohio.

Chapter 6

Applications of CPBS to Cancer Hazard Identification

6.1. INTRODUCTION

The field of genetic toxicology finds itself at a crossroads. On the one hand, the premise of the somatic mutation theory of cancer, which provides a scientific basis for the development of short-term tests for predicting cancers, has been amply vindicated by the discovery of oncogene activation. On the other hand, however, recent NTP-sponsored studies have cast doubt upon the performance of short-term tests as predictors of carcinogenicity (Tennant et al., 1987). Analysis of the NTP results by the CPBS shows that this is an incorrect conclusion resulting from an oversimplification (Rosenkranz and Ennever, 1988a). Also, it appears that we have no choice but to continue using short-term tests since the other alternatives are (a) not to test but to wait for untoward effects in our exposed human population and (b) to continue relying solely on animal bioassays.

If we choose the latter course, then we must fully realize that the concordance between rodents is only 70% (i.e., results on mice are only 70% predictive of results on rats and vice versa) and hence one cannot expect such an instrument to be more than 70% predictive of human cancers (Lave et al., 1988). There are, however, some additional facts that must be taken into consideration when considering the usefulness of short-term tests versus animal bioassays. These include the cost of the animal bioassay (about $1,000,000 per chemical) and its duration (up to three years). Since we can assume that the concordance between rodent bioassays and humans is not going to be better than 70% and if we realize that the proportion of carcinogens among chemicals in use is estimated to be 10%, then obviously we can expect a significant number of false-positive and false-negative cancer bioassay results. If we now include in our consideration a value judgment, say that the societal cost of a false negative is 10 times that of a false positive, the former resulting in disease and the latter resulting in a "safe" product that is being withheld from the market, then we can estimate that the standard animal bioassay is not a cost-effective tool for regulatory purposes and should only be used when widespread human exposure is anticipated and/or the projected sales either exceed the cost of testing or exceed the societal cost (Lave et al., 1988).

Given these considerations, the use of the short-term surrogate tests with all of their imperfections may constitute not only a cost-effective screening mechanism but also the only one capable of clearing the backlog of untested chemicals that are currently in commerce and industry. This number is estimated to be in excess of 70,000 chemicals. The fact that short-term tests may, under certain conditions, be better predictors of human carcinogens than animal bioassays (Ennever et al., 1987) gives an additional dimension to the problem and speaks for the continued use of short-term tests in identifying carcinogens.

Since the original publication of the CPBS (Chankong et al., 1985), we have gained much practical experience in using the method to select batteries of short-term tests predictive of carcinogenicity in animals, as well as to interpret the results of such tests. Moreover, the CPBS has now been used in a number of scientific as well as regulatory situations (Griesemer et al., 1985; Havel et al., 1985; Rosenkranz and Ennever, 1987, 1988a). In addition, in two international studies it correctly predicted the outcome (Ennever and Rosenkranz, 1986a; Rosenkranz et al., 1988). In the cancer hazard identification problem, it appears to be most applicable to decision making in the early phases of the product development process (Yander et al., 1987), and this latter feature will be elaborated herein.

In discussing the applicability of the CPBS, it must be noted that the predictions are greatly influenced by the data bases available as well as the reliability of the cancer bioassays against which the short-term tests are calibrated. Thus, using the most rigorous animal bioassay protocols currently available, those developed by the U.S. National Toxicology Program (NTP), in which lifetime exposures of mice and rats are carried out, it was found that mice are only 70% predictive of rats and vice versa (Gold et al., 1984; Tennant et al., 1987; Zeiger, 1987). As noted earlier, the concordance between the two groups with respect to carcinogenicity is only 70% and it would be unreasonable to expect rodent bioassays to be more than 70%

Applications of CPBS to Cancer Hazard Identification

predictive of human cancers. In addition, in view of the high cost of the animal carcinogenicity assay, most agents subjected to bioassays are those that have already been shown to give positive results in short-term tests. This has led to a sort of self-fulfilling prophecy, the result being that about 70% of chemicals tested in rodent cancer bioassays are found to be carcinogens (Tennant et al., 1987). Yet the common consensus holds that only approximately 10% of chemicals in commerce and industry are endowed with carcinogenic potential (Lave et al., 1988). This has resulted in a rather skewed data base, a situation which must be taken into consideration when evaluating the predictive power of batteries of short-term tests.

With respect to the results of short-term tests, some chemicals, particularly the potent mutagens and carcinogens, have been tested in most short-term tests because they were originally used for the development and "validation" of new methods. However, most other chemicals that are included in data bases have been tested in very few assays. The situation is complicated further by the fact that there are so many short-term tests. For example, a recent compilation by the International Agency for Research on Cancer lists approximately 150 short-term tests (IARC, 1988). Since many of the tests are, in fact, modifications of others, in order to obtain usable data it may be necessary to merge tests. This was done for the Gene-Tox compilation prepared under the aegis of the U.S. Environmental Protection Agency. It would, of course, be valuable if assays were carried out using a single protocol and if identical chemical batches were used for all assays. Data obtained under such rigorous conditions are being generated by NTP. However, in that program only a few assays are routinely included, and these may not be the most appropriate ones for battery deployment (Ennever and Rosenkranz, 1986b).

For illustrative purposes (especially for the characterization of test performance as exemplified by specificity and sensitivity), we shall for the present analysis use the Gene-Tox data base, fully aware of some of its limitations. However, the point stressed herein is that *the CPBS provides an objective approach to the interpretation of test results and for the deployment of batteries*. As data bases are expanded as well as improved, the CPBS predictions will also become even more reliable.

In order to demonstrate the CPBS, we will use some of the short-term test results, albeit not all, reported for acrylamide (Dearfield et al., 1988), a widely used industrial chemical. For some comparisons we shall also use data obtained with pentachloronitrobenzene (PCNB), a widely used pesticide. It should be noted that for the present illustrations we will not undertake an analysis of the reliability of the test results but will take the reported results at face value. Neither shall we examine the test protocols or the details of the procedures to assess the reliability of the interpretations.

Obviously, in practice when using the CPBS the quality of the test results must be ascertained. This requires the participation of adequately trained and experienced scientific and technical personnel.

The data base assembled for the acrylamide exercise is shown in Table 6.1.1. [It should be noted that while we selected nine assays for illustrative purposes, the chemical has in fact been tested in over 20 short-term tests (Dearfield *et al.*, 1988). It is most unlikely, however, when investigating a new chemical, that so many test results would be generated.] From an examination of Table 6.1.1 we see that of the nine test results, seven are positive and two negative. This is not an unusual situation—mixed results are often obtained. For the purpose of the present discussion, we assume that all of the results are reliable; that is, they are reproducible and show a dose dependence.

In previous analyses, which were based on published results, we indicated that when more than one result was reported for a particular assay, a majority rule was used. That is, a positive or negative result was assigned to a particular test based on whether the test gave a majority of positive or negative results, respectively, on the particular chemical, with the provision that if an equal number of positive and negative results were obtained, then the test was designated as positive (Havel *et al.*, 1985). However, in other situations, such as those that involve regulatory considerations (where it may be necessary to consider each individual test result), we have demonstrated that individual test results can be used in the CPBS, as long as each of the results is considered to be independent (Haimes *et al.*, 1987).

TABLE 6.1.1. Summary of Short-Term Test Results for Acrylamide

Abbreviation	Test name	Result	α^+	α^{-a}	Class[b]
Sty	Salmonella mutagenicity assay	−	0.612	0.806	III
Mly	Specific gene mutation in mouse lymphoma L5178Y cells	+	0.836	(0.5)	II (or I)
CHO	Specific gene mutations in Chinese Hamster ovary cells	+	0.781	(0.5)	II (or I)
Cvt	Mammalian cytogenetics *in vitro*	+	0.890	0.667	II
SCE	Sister chromatid exchanges	+	0.890	0.667	II
UDS	Unscheduled DNA synthesis	+	0.612	0.806	III
Drp	DNA damage assay using mammalian cells	+	0.890	(0.5)	II (or I)
Cbm	Bone marrow cytogenetics, *in vivo*	−	0.836	(0.5)	II (or I)
C3H	Transformation of C3H cells	+	0.890	(0.5)	II (or I)

[a] When α^- values could not be ascertained with significance, they were assigned a value of 0.5. Accordingly, some Class II tests might actually belong in Class I.
[b] Class I, $\alpha^+ \geq 0.8$ and $\alpha^- \geq 0.8$; Class II, $\alpha^+ \geq 0.8$ and $\alpha^- \leq 0.8$; Class III, $\alpha^+ \leq 0.8$ and $\alpha^- \geq 0.8$; Class IV, $\alpha^+ \leq 0.8$ and $\alpha^- \leq 0.8$.

Applications of CPBS to Cancer Hazard Identification

The CPBS has been discussed thoroughly in this volume as well as in a number of other publications (Chankong et al., 1985; Rosenkranz et al., 1986a). However, since the previous discussions were not problem specific we feel it necessary to reiterate some of the steps that are germane to the data analysis with respect to short-term tests. Briefly, the CPBS relates the sensitivity (α^+) and specificity (α^-) of an assay to the probability (θ^+) that the test chemical is a carcinogen based upon the experimental results. For the cancer hazard identification problem, α^+ and α^- are defined as

$$\alpha^+ = \frac{\text{Number of carcinogens found positive in assay}}{\text{Number of carcinogens tested}}$$

$$\alpha^- = \frac{\text{Number of noncarcinogens found negative in assay}}{\text{Number of noncarcinogens tested}}$$

For a single test, T_1, θ_1^+ is defined as

$$\theta_1^+ = \frac{\theta_0^+ \alpha_1^+}{\theta_0^+ \alpha_1^+ + (1 - \theta_0^+)(1 - \alpha_1^-)} \quad (6.1.1)$$

if the test result is positive, where θ_0^+ is the prior probability that the tested chemical is a carcinogen; and it is defined as

$$\theta_1^+ = \frac{\theta_0^+ (1 - \alpha_1^+)}{\theta_0^+ (1 - \alpha_1^+) + (1 - \theta_0^+)\alpha_1^-} \quad (6.1.2)$$

if the test result is negative.

The following equations can be used to sequentially calculate the results of batteries of tests as new tests T_i are added to the battery:

$$\theta_i^+ = \frac{\theta_{i-1}^+ \alpha_i^+}{\theta_{i-1}^+ \alpha_i^+ + (1 - \theta_{i-1}^+)(1 - \alpha_i^-)} \quad (6.1.3)$$

if the test result T_i is positive, and

$$\theta_i^+ = \frac{\theta_{i-1}^+ (1 - \alpha_i^+)}{\theta_{i-1}^+ (1 - \alpha_i^+) + (1 - \theta_{i-1}^+)\alpha_i^-} \quad (6.1.4)$$

if the test result T_i is negative. Note that the final value of θ^+ is independent of the order in which the test results are calculated. Also note that the form of Eqs. (6.1.3) and (6.1.4) is based on the independence of the test results, which is assumed to hold throughout the following discussions.

As noted earlier, the quantity θ_0^+ is the prior probability of carcinogenicity. The value of θ_0^+ can be based on expert intuition. For example, the presence of two hydrogen atoms on the carbon adjacent to the NNO function in nitrosamines may indicate, with a high probability, that the chemical is endowed with carcinogenic potential. Similarly, the presence of a bay region has also been associated with the carcinogenicity of polycyclic aromatic hydrocarbons (Jerina et al., 1976), as well as the ability of an agent to be metabolized to an electrophilic intermediate (Miller, 1970; Miller and Miller, 1977). Alternatively, θ_0^+ might be the probability derived from an expert system to predict activity based upon structural considerations (see Rosenkranz and Klopman, 1987; Rosenkranz et al., 1986b) or, finally, it could reflect the prevalence of carcinogens in the universe of chemicals [e.g., we might assume, based upon a large-scale screening study conducted in Japan, that 10% of chemicals are endowed with a carcinogenic potential (Matsushima, 1987)]. For analytical purposes we have assumed a value for θ_0^+ of 0.50, i.e., an equal chance that the chemical is or is not a carcinogen. It has been shown previously that the assignment of θ_0^+ does not affect the final θ^+ value from a qualitative viewpoint. The essential (qualitative) feature of the prediction is the increase in knowledge gained by considering the testing results, rather than the actual value of θ^+ itself (Ennever and Rosenkranz, 1986c). For example, assuming θ_0^+ is 0.50 and finding that θ^+ is 0.93 after testing represents a significant gain in knowledge while going from 0.50 to 0.53 might not be an acceptable increase in knowledge. In addition, batteries should be designed so that θ^+ is nearly independent of θ_0^+ over a considerable range of θ_0^+ values (this property is designated as a robust prediction; see below).

Currently, the values for α^+ and α^- used in our examples are derived from the augmented Gene-Tox data base improved by cluster analysis (Palajda and Rosenkranz, 1985; Pet-Edwards et al., 1985a, b; see also Chapter 3 of this volume). When a value of α^- could not be estimated owing to a paucity of data on noncarcinogens, it was assigned a noninformative value of 0.5 (Chankong et al., 1985). The calculations of θ^+ and distributions of θ^+ as a function of θ_0^+ were carried out by a program designated CPBS^{TM}-1 that operates on an IBM personal computer.

6.2. AN ILLUSTRATION OF CPBS PREDICTION ON ACRYLAMIDE

The sensitivities and specificities of the assays are listed in Table 6.1.1, as are the classes to which they belong. Even though the CPBS is available in computerized form, it would be instructive to go through some of the calculations on acrylamide in a step-by-step fashion. If we take the first test

Applications of CPBS to Cancer Hazard Identification

result listed, Sty, which is negative (see Table 6.1.1) and use Eq. (6.1.2) with a prior probability of 0.50, we have

$$\theta_1^+ = \frac{(0.5)(1 - 0.612)}{(0.5)(1 - 0.612) + (1 - 0.5)(1 - 0.806)} = 0.3249$$

This implies that a negative result in Sty carries with it a 32.5% chance of carcinogenicity or a 67.5% chance of noncarcinogenicity—a change of 17.5% over the original probability of 0.5.

[It should be noted that positive and negative results in Sty do not carry equal weights. This is a consequence of the relatively low sensitivity and high specificity of the Sty assay. A test with such characteristics is designated as a Class III test (see the legend below Table 6.1.1). Had Sty given rise to a positive result, we would have used Eq. (6.1.1) and would have found that

$$\theta_1^+ = \frac{(0.5)(0.612)}{(0.5)(0.612) + (1 - 0.5)(1 - 0.806)} = 0.7597$$

In other words, a positive result in Sty would have carried with it a 76.0% chance of carcinogenicity compared to the original probability of 0.50.]

Continuing with the results listed in Table 6.1.1, the second test result for acrylamide is a positive result in Cvt. We now use Eq. (6.1.3) with $i = 2$. The value of θ_1^+ is 0.3249 and the values of α_2^+ and α_2^- are the sensitivity and specificity, respectively, of the test Cvt. The computation of the probability that acrylamide is a carcinogen based on the two test results would be given as

$$\theta_2^+ = \frac{(0.3249)(0.890)}{(0.3249)(0.890) + (1 - 0.3249)(1 - 0.667)} = 0.5625$$

We continue with the computations for the remaining seven tests in the same manner, using Eq. (6.1.3) each time a test result is positive and using Eq. (6.1.4) when the test is negative. When computing θ_n^+ for the nth test, we use the result from the previous step, θ_{n-1}^+, along with the sensitivity and specificity of the nth test in the appropriate equation [i.e., either Eq. (6.1.3) or (6.1.4)]. A summary of the results on acrylamide given as a sequence of nine computations is presented in Table 6.2.1. The calculations show a clear indication that acrylamide is predicted to be a carcinogen—the

TABLE 6.2.1. CPBS Prediction of the Carcinogenicity of Acrylamide

	Battery[a]	θ^+
θ_0^+	—	0.500
θ_1^+	Sty	0.3249
θ_2^+	Sty + Cvt	0.5625
θ_3^+	Sty + Cvt + UDS	0.8026
θ_4^+	Sty + Cvt + UDS + SCE	0.9157
θ_5^+	Sty + Cvt + UDS + SCE + Cbm	0.7810
θ_6^+	Sty + Cvt + UDS + SCE + Cbm + CHO	0.8479
θ_7^+	Sty + Cvt + UDS + SCE + Cbm + CHO + Mly	0.9031
θ_8^+	Sty + Cvt + UDS + SCE + Cbm + CHO + Mly + Drp	0.9432
θ_9^+	Sty + Cvt + UDS + SCE + Cbm + CHO + Mly + Drp + C3H	0.9673

[a] Note that the computation procedure using Bayes' equation is commutative; i.e., the same final result will be obtained irrespective of the order in which the test results were computed.

probability of carcinogenicity for this chemical is 0.9673, even though two of the test results were negative, that is, Sty and Cbm.

In previous studies it was shown that for preliminary classification, when starting with a prior probability of 0.5, θ^+ values of ≥ 0.7 are taken as indications of carcinogenicity while θ^+ values of ≤ 0.3 are an indication of noncarcinogenicity. Values of θ^+ between 0.3 and 0.7 are considered to be inconclusive and may require additional testing (Ennever and Rosenkranz, 1986b). It must be pointed out, however, that during the development of consumer products, one might want to be much more risk-averse and require much lower probabilities (e.g., θ^+ values of ≤ 0.1) before considering a product for further development (this will be discussed in more detail later).

In order to arrive at a probability of carcinogenicity of 0.967 for acrylamide, we used nine test results at an estimated cost of $82.7K (see Battery 1 in Table 6.2.2). Yet, an examination of the incremental prediction (Table 6.2.1) suggests that θ^+ values of ≥ 0.8 can be obtained by using batteries consisting of three or four tests. Thus batteries consisting of fewer than nine tests that are still predictive and yet more cost-effective could be found. This will be discussed in more detail in Section 6.3.

It must be remembered that CPBS predictions, such as those listed in Tables 6.2.1 and 6.2.2, predict neither actual carcinogenicity nor carcinogenic potency, but rather they identify a hazard. Whether or not this hazard is realized depends upon many factors such as dose and duration of exposure, mode of administration, the metabolism and detoxification of the test chemical, the host's innate immunological and homeostatic defense mechanisms, its ability to repair DNA, and its subsequent exposure to promotional agents as well as other less well-defined factors, such as diet and lifestyle.

Applications of CPBS to Cancer Hazard Identification

TABLE 6.2.2. Summary of CPBS Predictions for Acrylamide

Battery	Composition	θ^+	Approximate cost in 10^{-3} U.S. dollars
1	Sty, Cvt, UDS, SCE, Cbm, CHO, Mly, Drp, C3H	0.9673	82.7
	Semi-ideal		
2	Sty, Cvt, UDS, SCE	0.9157	23.7
3	Sty, UDS, Drp, C3H	0.8284	27.3
	Biological or regulatory relevance		
4	Mly, Cbm, Cvt, C3H	0.7231	43.5
5	Mly, Drp, Cvt, SCE	0.9340	34.9
6	Sty, Cbm, Cvt, C3H	0.4292	36.8
7	Sty, Cvt, Drp, C3H	0.8031	26.3
8	Mly, Cvt, Drp, C3H	0.9340	33.0
	Risk averse		
9	SCE, Cvt, Drp	0.9271	26.4
10	Sty, Cvt, SCE	0.7745	16.2
11	Sty, Cvt, SCE, Drp	0.8595	28.2
12	Sty, Cvt, SCE, Mly	0.8517	24.7

One of the advantages of using the computerized version of CPBS is the possibility of obtaining complete distributions of predicted carcinogenicity, i.e., θ^+ values as a function of the prior probability, θ_0^+, values. This enables rapid modeling of a number of situations and also allows a better understanding of the battery performance and the influence of individual test results on θ^+.

The effect of each of the nine test results (given in the order listed in Table 6.1.1) on the probability of carcinogenicity θ^+ as a function of θ_0^+ is shown in Fig. 6.2.1. Thus, for the two tests, Sty and Cvt, the former being negative and the latter positive, at a prior probability of 0.5, the earlier calculations had already shown that the predicted probability (θ^+ value) of this battery is 0.56 (θ_2^+, Table 6.2.1). This, however, is not a significant increase in knowledge after testing when compared to a θ_0^+ of 0.5. Such a prediction is not much better than the toss of a coin. The complete distribution of these two test results is given by the second curve from the bottom of Fig. 6.2.1. It indicates that irrespective of the value of the prior probability θ_0^+, the final θ^+ value is not affected very much by the two test results. In other words, when Sty is negative and Cvt is positive, the predicted probability is approximately the same as the prior probability. In this case, the curve is almost diagonal, which indicates no increase in knowledge after testing.

On the other hand, for the prediction consisting of all nine test results (the top curve of Fig. 6.2.1), there is very little effect of the prior probability

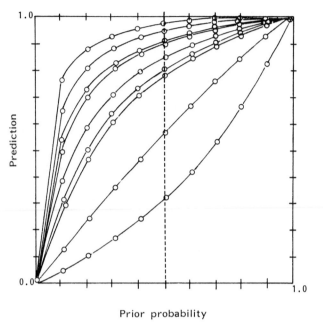

FIGURE 6.2.1. The predicted carcinogenicity of acrylamide as a function of the prior probability (θ_0). The curves correspond to the tests enumerated in Table 6.2.1.

θ_0^+ on the prediction θ^+ over the range $0.2 < \theta_0^+ < 0.95$. This is the hallmark of a robust prediction.

6.3. BATTERY SELECTION CONSIDERATIONS

As can be seen from even a cursory examination of Table 6.2.1 and Fig. 6.2.1, it would seem to be uneconomical to run as many as nine tests to predict the carcinogenicity of acrylamide since even the results of a three-test battery can indicate a significant probability of carcinogenicity. The problem then becomes one of battery selection.

As mentioned earlier, a recent compilation of the International Agency for Research on Cancer (IARC, 1988) lists approximately 150 tests. If we decided that a battery should consist of four tests, and if we are not more selective as to which of the four tests should be included, this would yield 20,260,275 possible combinations to choose from. Similarly, if we selected from among the 40 tests that we have abstracted from the Gene-Tox data base, this would yield 91,390 combinations consisting of four tests.

Obviously, there must be a further reduction in the number of possible batteries that can be selected.

One such solution is offered using the multiple objective dynamic programming approach given in Chapter 4 (see also Pet-Edwards, 1986; and an approach by Hu, 1985) in which testing cost and battery predictivities are determinants. An example of this method is given in Chapter 4. This approach can greatly reduce the number of tests that must be considered. However, even though cost is undoubtedly of great importance in the cancer hazard identification problem, the approach based on dynamic programming does not take into consideration the biological relevance of the tests that are combined into a battery. In addition, the availability of some tests is not universal. For example, such tests as DRL (Drosophila recessive lethals) or SHE (Syrian hamster embryo transformation) may require special expertise and/or facilities, and this can add both additional costs and an unacceptable delay in obtaining test results. Finally, there are tests that have not been sufficiently validated or tests that have been restricted to only certain chemical classes. Such tests should not be included in a battery either, since their performance characteristics are unknown. (It should be mentioned that heretofore we have restricted our characterization of tests to the peer-reviewed Gene-Tox compilation, which by and large was completed in 1981, and to the NTP-generated data, which include fewer tests. Obviously, both in the published literature as well as in the files of regulatory agencies and industry, there are many more test results that have not been included in the assessment of test performance.)

Given these restrictions, the selection of tests for inclusion in a battery is not entirely algorithmic and requires considerable thoughtful scientific input. In the developmental stage of CPBS, we identified an "ideal battery" (Chankong *et al.*, 1985) as one consisting of as many Class I tests (high sensitivity and high specificity) as are available, and an equal number of Class II tests (high sensitivity and low specificity) and Class III tests (low sensitivity and high specificity). This was based on considerations regarding the complementarity of test performances in Bayes' theorem (see Section 4.3). Another recommendation was that the number of Class I tests be odd, in order to allow the use of the majority rule (see Chankong *et al.*, 1985). However, since the interpretation of test results is not necessarily by majority rule but can be based on using Bayes' theorem (e.g., predictivities), we do not have to abide by the requirement of an odd number of Class I tests.

An examination of the Gene-Tox data base indicates that, in practice, there are very few feasible Class I tests available. We originally identified BHK (baby hamster kidney transformation) and DRL as Class I assays. However, owing to a lack of inter- and intralaboratory reproducibility, BHK is no longer used for screening purposes. The DRL assay, in addition to being expensive, also requires special expertise that may not be generally available. With respect to the test results on acrylamide (Table 6.1.1), it should be mentioned that no Class I tests are included.

As a further refinement of the CPBS method, we defined a "semi-ideal" battery as one consisting of two Class II and two Class III tests (Ennever and Rosenkranz, 1986c; Rosenkranz and Ennever, 1988a, b; Rosenkranz *et al.*, 1986a). Within the context of the data base on acrylamide, several such combinations of assays can be identified. For illustrative purposes, two will be analyzed further. (Note that in the ensuing discussion, we are using a prior probability of 0.5 in our computations.) One of these batteries could consist of Sty, Cvt, UDS, and SCE—denoted as Battery 2. This battery gives a θ^+ value of 0.9157, i.e., a 91.6% chance of the carcinogenicity of acrylamide (see θ_1^+ to θ_4^+ in Table 6.2.1 and Battery 2 in Table 6.2.2). Since most companies would be risk averse, they would not tend to differentiate between a 96.7% of cancer predicted by the nine-test battery (θ^+) of Table 6.2.1 at an approximate cost of \$83K and a 91.6% chance derived from a four-test battery costing \$24K (Battery 2, Table 6.2.2). Both results would indicate a relatively strong potential for carcinogenicity. Note that both batteries provide essentially the same type of usable information, but their costs are quite different.

The distribution of predicted probabilities as a function of θ_0^+ values for Battery 2 (Fig. 6.3.1) shows that irrespective of the assumed θ_0^+ value, there is a considerable gain in information. For example, when θ_0^+ is 0.1, θ^+ becomes 0.55; while for θ_0^+ of 0.5 we have a θ^+ of 0.92; and for θ_0^+ of 0.70, $\theta^+ = 0.93$. This is further exemplified by the convexity of the curve in Fig. 6.3.1, which is an indication of a low amount of influence of θ_0^+ on θ^+ between values of θ_0^+ between 0.2 and 0.9. This again indicates a robust prediction.

Obviously, a variety of "semi-ideal" batteries can be constructed (another one would be Sty, UDS, Cvt, and Drp), but for the purpose of our discussion let us examine a battery consisting of Sty, UDS, Drp, and C3H. While Sty and UDS are definitely Class III tests, Drp and C3H, because of paucity in the data regarding noncarcinogens, have been assigned specificity values of 0.5, which places them for predictive purposes in Class II. That makes the following analysis a "worst case" analysis, as conceivably the two tests could belong to Class I.

Using this "worst case" analysis and applying CPBS to the test results, we still come up with an 82.8% chance of carcinogenicity (Battery 3 of Table 6.2.2). This is lower than for the previous battery (Battery 2), but is still predictive of carcinogenicity. However, even though there was a definite increase of information as a result of testing, the shape of the curve for Battery 3 is not as robust as for Battery 2 (compare Figs. 6.3.1 and 6.3.2).

Applications of CPBS to Cancer Hazard Identification

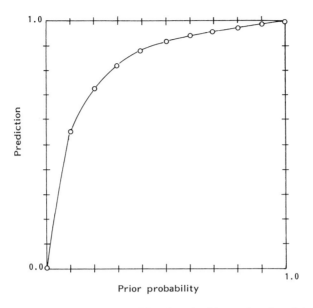

FIGURE 6.3.1. The predicted carcinogenicity of acrylamide as a function of θ_0 based upon the results of a battery consisting of Sty, UDS, Cvt, and SCE.

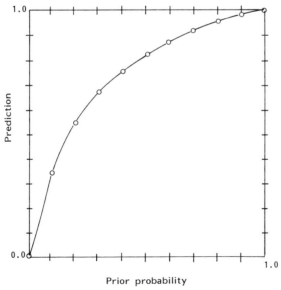

FIGURE 6.3.2. The predicted carcinogenicity of acrylamide as a function of θ_0 based upon the results of a battery consisting of Sty, UDS, Drp, and C3H.

In addition, the cost of Battery 3 is $4K more than the cost of Battery 2 (see Table 6.2.2). Accordingly, in terms of the uncertainties regarding the specificities of two of the tests in Battery 3, the cost of Battery 3, and the robustness of the prediction given by Battery 3, it would appear that Battery 2 is superior in terms of cost and reliability.

In addition to setting up a battery using the preceding cost and performance considerations, the problem could also be approached from a biological or a regulatory point of view. A number of suggestions have been made that a battery should consist of tests varying in phylogenetic complexity as well as in the genetic end points included (Weisburger and Williams, 1981). Special emphasis has been placed by some on the importance that both *in vitro* and *in vivo* assays be included (Ashby, 1986). Within the present context of the acrylamide data base, a battery consisting of Mly (gene mutation/chromosomal mutations in cultured mammalian cells), Cvt (cytogenetics in cultured mammalian cells), Cbm (*in vivo* bone marrow cytogenetics), and C3H (oncogenic cell transformations) could be considered. Such a battery encompasses a broad spectrum of end points and phylogenetic systems. When we apply the CPBS to this combination of tests, a θ^+ of 0.7231 is obtained—a 72.3% chance of carcinogenicity (Battery 4, Table 6.2.2). A somewhat similar battery can be constructed in which Drp, an *in vivo* assay, replaces Cbm, another *in vivo* test. This battery predicts carcinogenicity with a 0.955 probability (Battery 5, Table 6.2.2). The costs of the last two batteries are $43.5K and $34.9K, respectively.

Since the Sty assay is so widely used, one might question a battery that does not include this assay. Thus, let us examine an alternative battery consisting of Sty, Cbm, Cvt, and C3H. This battery yields a θ^+ of 0.4293 that is clearly inconclusive (Battery 6, Table 6.2.2 and Fig. 6.3.3) for the results obtained. In such a situation one would probably want to run additional tests. This is clearly a case where mixed results (in this case two positive and two negative results) have made the prediction inclusive. We will later demonstrate how one can construct batteries of tests such that the possibility of inconclusive results would be minimized.

Another biologically relevant battery that includes Sty also predicts the carcinogenicity of acrylamide with a fairly high probability (see Battery 7 in Table 6.2.2 and Fig. 6.3.4). Note that this battery, consisting of Sty, Cvt, Drp, and C3H, is a relatively low-cost battery (i.e., $26.3K). If we wish to diversify the battery further and substitute Mly for Sty (a Class II or I test for a Class III), this results in an increase in the certainty that acrylamide is a carcinogen because we have replaced a negative test result with a positive result (Battery 8, Table 6.2.2). All of the biologically relevant batteries (with the exception of Battery 6) predict that acrylamide is carcinogenic (see Batteries 4, 5, 7, 8, in Table 6.2.2).

Applications of CPBS to Cancer Hazard Identification

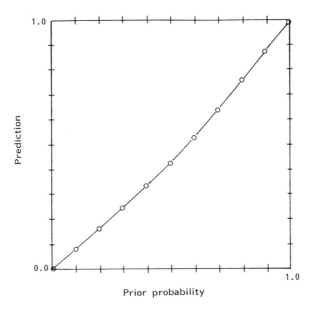

FIGURE 6.3.3. Inconclusive prediction of the carcinogenicity of acrylamide as a function of θ_0 based upon the results of a battery consisting of Sty, Cbm, Cvt, and C3H. Note that the curve is almost diagonal, which indicates little increase in knowledge after testing.

In the preceding discussion we examined the results of eight different batteries (based on a variety of selection criterion) on acrylamide. From the analysis it appears that regardless of the rules used to construct the battery, acrylamide is very likely carcinogenic. In fact, acrylamide has been reported to be carcinogenic to experimental animals (IARC, 1986). Had acrylamide been a new chemical currently in the product development stage, management might consider looking for an alternate product or putting into place a variety of containment measures.

Normally, in the development of a new chemical, one does not have a set of assay results already available. Instead one must decide what tests should be conducted on the new chemical. In this case the past performances of the tests on other chemicals should be considered, as well as the performance characteristics of the various combinations of tests and the various possible outcomes of the batteries of tests. We noted earlier that "ideal" batteries are often difficult to construct owing to the general unavailability of Class I tests and the high costs of the few available ones. In addition, we know that each test is characterized by a pattern of false positive and false negative results (indeed, the recognition of this latter principle is the basis for using Bayes' theorem). For these reasons, one might want to construct batteries that will reflect the purpose of the testing program.

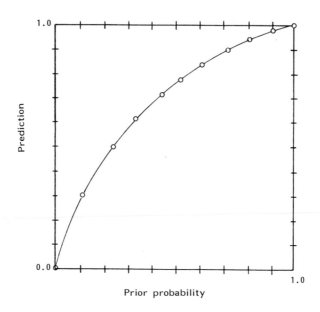

FIGURE 6.3.4. The predicted carcinogenicity of acrylamide as a function of θ_0 based upon the results of a battery consisting of Sty, Cvt, Drp, and C3H.

Regulatory agencies are charged with protecting the public welfare. Since they must prioritize their activities, they have as their mission the identification of the most noxious of the agents found in the environment and the workplace. Thus they would be concerned with minimizing the number of false positives (i.e., noncarcinogens falsely predicted to be carcinogenic), even if this is accomplished at the cost of getting a number of false negatives (i.e., chemicals that are erroneously predicted to be noncarcinogenic even though they are in fact carcinogens). To accomplish this, a battery consisting of tests that are highly specific but not necessarily highly sensitive may be used (e.g., Class III tests, such as Sty and UDS). On the other hand, manufacturers of consumer products are primarily interested in bringing safe chemicals to the market. Therefore, they would want to minimize the number of false negatives, and accept the possibility of false positives (that is they would want to avoid the possibility of failing to identify that a chemical that is carcinogenic, with the possibility that some noncarcinogens might be falsely predicted to be carcinogenic). Accordingly, in order to accomplish this, batteries that are reliable in predicting noncarcinogenicity are required. To achieve this type of performance one would select batteries that are highly sensitive, but not necessarily highly specific (e.g., Class II assays) (Ennever and Rosenkranz, 1986c). In

Applications of CPBS to Cancer Hazard Identification

addition, more stringent risk-averse criteria can be required as well. For example, one might require that θ^+ be less than 0.1 or 0.05 in order to be fairly certain that the product is not a cancer hazard.

In our example with acrylamide, suppose we would like to minimize false negatives and would not continue the development of the chemical if θ^+ is greater than 0.3. Such a risk-averse battery could consist of SCE, Cvt, and Drp. This combination predicts a 0.927 probability that acrylamide is carcinogenic (Battery 9, Table 6.2.2). In addition, the distribution of θ^+ as a function of θ_0^+ is also very robust (see Fig. 6.3.5). It should be noted that in this last calculation we assumed Drp to be a Class II test. In fact, it could be a Class I test, which would increase the reliability of the prediction further (by reducing the possibility of getting false positives). Obviously, an additional test such as Mly (which can be taken as a Class II test) would increase the probability of carcinogenicity to 0.934 (i.e., Battery 5, Table 6.2.2) if it gave a positive result. However, it is questionable whether such an additional test is warranted, since in a risk-averse situation a chemical with such a high probability of carcinogenicity will either not be developed or its use will be restricted to controlled environments. Note that even if Mly resulted in a negative result, this would not bring θ^+ down to the acceptable level (i.e., below 0.3).

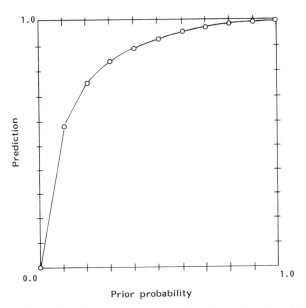

FIGURE 6.3.5. The predicted carcinogenicity of acrylamide as a function of θ_0 based upon the results of a battery consisting of SCE, Cvt, and Drp.

As has been mentioned on several occasions already, since Sty is almost universally used, one might consider including this assay in a risk-averse battery even though Sty is a Class III test. Thus, we may construct a battery consisting of Sty, Cvt, and SCE that predicts with a probability of 0.775 that acrylamide is carcinogenic (Battery 10, Table 6.2.2). However, this prediction shows considerable dependence on the θ_0^+ values (see Fig. 6.3.6). Thus, one might want to include a fourth test such as either Drp (Battery 11, Table 6.2.2) or Mly (Battery 12, Table 6.2.2), which predict the carcinogenic potential of acrylamide to be 0.8595 and 0.8517, respectively. Note that some of the batteries that were constructed based upon biological relevance are also risk-averse batteries—see for example Battery 8, Table 6.2.2.

Class II tests are considered to be risk averse because they give a good predictivity for negative tests results (that is they give few false negatives). Basically, in using risk-averse batteries, the question is how many negative results "assure" the noncarcinogenicity of the test chemical. Hence, for illustrative purposes it would be more instructive to consider a chemical that gave primarily negative results in the assays. Accordingly, for explanatory purposes we will examine another chemical—pentachloronitrobenzene.

Pentachloronitrobenzene (PCNB) is a widely used pesticide. A search of the Gene-Tox data base (Palajda and Rosenkranz, 1985) indicates that

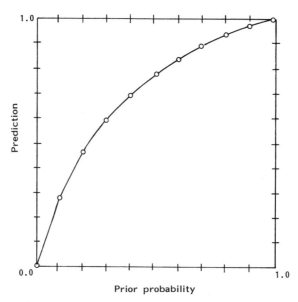

FIGURE 6.3.6. The predicted carcinogenicity of acrylamide as a function of θ_0 based upon the results of a battery consisting of Sty, Cvt, and SCE.

Applications of CPBS to Cancer Hazard Identification

TABLE 6.3.1. Summary of Short-Term Results for Pentachloronitrobenzene

Abbreviation	Test name	Result	α^+	α^+	Class
Sty	Salmonella mutagenicity assay	−	0.612	0.806	III
UDS	Unscheduled DNA synthesis	−	0.612	0.806	III
Cvt	Mammalian cytogenetics *in vitro*	+	0.809	0.667	II
SCE	Sister chromatid exchanges	−	0.890	0.667	II
HMA	Host-mediated assay	−	0.767	0.800	III

it has been tested in five systems resulting in four negatives and one positive (Table 6.3.1). Again, for the reasons outlined earlier, all of the batteries constructed for the present illustrative purposes include Sty. If we use all of the test results (Battery 1, Table 6.3.2) we arrive at a θ^+ value of 0.030, which is equivalent to a 97% chance that PCNB is noncarcinogenic. Moreover, there is very little effect of θ_0^+ on θ^+ (see Fig. 6.3.7). A "semi-ideal" battery (Battery 2 of Table 6.3.2) shows a θ^+ value of 0.0924. And, finally, a risk-averse battery (Battery 3 of Table 6.3.2) yields a value of 0.1745 for θ^+. It should be noted, however, that Battery 3 consists of only three tests, one of which gave a positive result. For more definitive results, we might want to include a fourth test. For the present example, we have included HMA in the risk-averse battery because its sensitivity is high enough that it could almost be classified as a Class I test. Inclusion of HMA results in a θ^+ value of 0.0604 (Battery 4, Table 6.3.2), which indicates a 94.0% chance that PCNB is not carcinogenic.

If we use the criterion that θ^+ values that are less than 0.3 are indicative of noncarcinogenicity, then PCNB would be classified as a noncarcinogen by all four batteries. Were we to use a more stringent criterion that might require θ^+ values of ≤ 0.1, then this chemical would also qualify as a noncarcinogen for all batteries except for Battery 3. However, if the criterion for acceptance is θ^+ values of ≤ 0.05, then the results are inconclusive, and further testing might be warranted. [In fact, PCNB has been tested in animals and has been shown to be noncarcinogenic (Nesnow *et al.*, 1987).]

TABLE 6.3.2. Summary of CPBS Predictions for Pentachloronitrobenzene

Battery	Description	Composition	θ^+
1	All tests	Sty, UDS, Cvt, SCE, HMA	0.0300
2	Semi-ideal	Sty, UDS, SCE, Cvt	0.0924
3	Risk averse	Sty, Cvt, SCE	0.1745
4	Risk averse	Sty, Cvt, SCE, HMA	0.0604

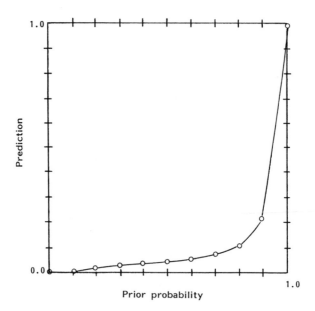

FIGURE 6.3.7. The predicted lack of carcinogenicity of pentachloronitrobenzene as a function of θ_0 based upon all of the test results (Sty, UDS, Cvt, SCE, and HMA).

6.4. CPBS AND DECISION ANALYSIS

The data presented thus far describe the use of the CPBS for interpreting test results. Obviously, unless the battery of tests includes some tests that are unreliable, the rational approach would be to use all of the available test results to calculate θ^+. On the other hand, analysis of the data reveals that abbreviated batteries consisting of three or four tests are also sufficiently predictive and that battery assembly rules could be devised to select combinations of tests that would both be cost effective and have a biological basis.

Suppose that a series of tests has been agreed upon, or a variable battery selected from a fixed panel of tests is being considered; one must then have a plan of action regarding the possible test results. For example, a battery of three tests can give rise to eight different results, and four tests can give rise to 16 different results. Before embarking upon a testing program, one must have rules regarding the interpretation of the results and what further testing may be required. Thus, let us examine the situation of a three-test battery. Suppose that a chemical company decides *a priori* that a θ^+ value of ≥ 0.7 represents an unacceptable potential hazard and that such a chemical will be rejected without further testing or development. On

Applications of CPBS to Cancer Hazard Identification

TABLE 6.4.1. Predicting the Possible Results of a Battery Consisting of DRL, Cvt, and UDS

	DRL	Cvt	UDS	θ^+
1	−	−	−	0.016
2	−	−	+	0.096
3	−	+	−	0.207
4	+	−	−	0.255
5	−	+	+	0.632
6	+	−	+	0.692
7	+	+	−	0.847
8	+	+	+	0.973

the other hand, suppose a value of ≤0.3 indicates a lack of potential carcinogenicity, i.e., an acceptable risk; and values between 0.3 and 0.7 are inconclusive and the chemical would require further testing. If our battery were to consist of DRL, Cvt, and UDS, there would be eight possible results for the battery—it could yield three results that indicate (by the above criteria) a potential for carcinogenicity (Combinations 6-8, Table 6.4.1), four results that suggest noncarcinogenicity (Combinations 1-4, Table 6.4.1), and only one result (Combination 3) that is inconclusive. This is a very definitive battery, which derives from the fact that it is an "ideal" battery consisting of a Class I, Class II, and Class III test.

Now consider the battery that consists of Sty, Mly, and SCE (Table 6.4.2) (two Class II tests and one Class III test). In this situation we have three inconclusive results if we use the cutoffs of 0.3 and 0.7. For the cases where the results exceed 0.7 or fall below 0.3, we already have decided on a plan of action—namely, rejecting or accepting the product, respectively.

TABLE 6.4.2. Effect of Fourth Test on a Three-Test Battery Consisting of Sty, Mly, and SCE

	Sty	Mly	SCE	θ^+	θ^+ with CHO⁻	Mnt⁻	Cvt⁻	UDS⁻
1	−	−	−	0.0253				
2	−	+	−	0.1168	0.0547	0.0416	0.0213	0.0599
3	+	−	−	0.1458	0.0695	0.0531	0.0273	0.0759
4	−	−	+	0.2969	0.1559	0.1218	0.0649	0.1689
5	+	+	−	0.4649				
6	−	+	+	0.6825				
7	+	−	+	0.7350				
8	+	+	+	0.9339				

We must, however, also have a plan of action regarding the inconclusive results. This may involve additional testing. The question then is, which test should be included as a fourth test?

Let us return to the inconclusive results obtained with the ideal battery, DRL, Cvt, and UDS (Combination 5, Table 6.4.1). Suppose that we are considering either Sty or SCE as the fourth test. Modeling the above situation with Sty as the fourth test, we see that a positive result with Sty indicates carcinogenicity, while a negative result still leaves θ^+ in the inconclusive range, $\theta^+ = 0.4529$ (see Table 6.4.3). On the other hand, when SCE is the fourth test, both a positive result ($\theta^+ = 0.8212$) and a negative result ($\theta^+ = 0.2204$) are conclusive (Table 6.4.3). Hence, if the choice were between Sty and SCE, obviously SCE should be the test selected if we want to ensure that the results will be conclusive. (In practice, considerations of testing costs and the availability of the tests may also be important in the selection of the "best" next test.) Analyses of the preceding type can, of course, be carried out with a number of candidate tests.

Let us now consider a different problem. Suppose that we want an initial three-test battery, to be selected from a panel of seven tests (Sty, Mly, CHO, Mnt, Cvt, SCE, and UDS), and that the initial battery must include Sty (Yander *et al.*, 1987). In addition, suppose we require that if the initial three tests are uniformly negative and θ^+ is ≤ 0.1, no further testing is required and the chemical is considered acceptable. If, however, θ^+ is ≥ 0.1 or if one of the test results is positive, then a fourth test is to be selected from among the remaining panel. In order for the chemical to be acceptable, the fourth test must also be negative and θ^+ for the four-test battery must be ≤ 0.1. If this were the situation, then there are 120 possible results stemming from the 15 possible three-test batteries that include Sty. In addition, each of the 15 batteries can yield three possible combinations of one positive and two negative results. This means that there will be 45 possibilities which will require a fourth test (Ennever and Rosenkranz, 1987b). Analysis also indicates that the choice of the tests included in the initial battery as well as the additional test must be considered carefully.

TABLE 6.4.3. Effect of Fourth Test on a Three-Test Battery Consisting of DRL, Cvt, and UDS

DRL	Cvt	UDS	θ^+	Fourth test	θ^+ After fourth test	
					−	+
−	+	+	0.6323	Sty	0.4529	0.8447
				SCE	0.2204	0.8212

Let us examine one of the combinations, namely, Sty, Mly, and SCE, that was considered earlier (Table 6.4.2). By the new, more stringent criteria, Combination 1, i.e., three negatives, yields a θ^+ value of 0.0253 and clearly meets the criteria for acceptance without further testing at this stage. Furthermore, analysis of the predicted θ^+ values indicates that when Combinations 2, 3, and 4 (which give one positive and two negative results) are obtained, the chemical should be tested further using either UDS, Cvt, CHO, or Mnt. Recall that the requirement for the fourth test is that if it is negative, then the predicted θ^+ value for the total test battery should be ≤ 0.1. Obviously, the selection of the fourth test is pivotal. Inclusion of Cvt appears to satisfy the requirements, since all three test result combinations (i.e., 2, 3, and 4) that include the negative result in Cvt meet the criterion of $\theta^+ \leq 0.1$ (see Table 6.4.2). Note that Cvt is a Class II test (with a high sensitivity and an acceptable specificity), which makes it a good test for predicting negative results. The same type of analysis can be used to investigate each of the remaining 14 different batteries composed of three tests, but will not be done here.

In addition to being able to make such *a priori* predictions, we can also estimate the likelihood that these predictions will occur. This information is contained in the denominator of Bayes' equation [Eqs. (6.1.1)-(6.1.4)] (Ennever and Rosenkranz, 1986c, 1988b). Thus, for example, for the earlier example using DRL, Cvt, and UDS (Table 6.4.1), we can estimate how often each of the combinations of eight test results is likely to occur (see Fig. 6.4.1). Note that the distribution of predictions is quite symmetrical. This is a direct consequence of the fact that this battery is an "ideal" one consisting of a Class I, Class II, and Class III test. Such a combination is equally good at predicting both carcinogens and noncarcinogens.

On the other hand, if we look at the battery consisting of Sty, Mly, and SCE (Table 6.4.2) and the predicted occurrence of the test results (illustrated in Fig. 6.4.2), we find that the distribution is not symmetrical. This is a reflection of the inclusion of Class II tests. Note that the predictions for the negative results are generally more reliable than the predictions for the positive results and that we would expect the results for this battery to be inconclusive a fairly high proportion of the time.

Knowledge of the expected occurrence of test results can facilitate decisions as to whether a test product should be tested further or whether it should be dropped from further consideration when a certain series of test results has been obtained. This analysis is especially valuable in sequential testing. For example, suppose the first test conducted was Sty, and suppose the test gave a positive result. This test is known to give at least some false positive results (i.e., it is a Class III test), so one could then estimate the likelihood of getting, for example, three negative results in

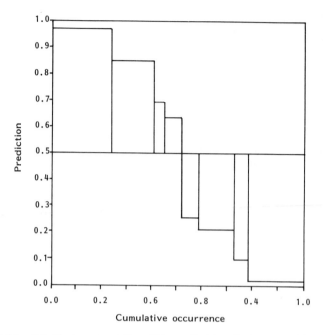

FIGURE 6.4.1. Graphical representation of the predictivity of a battery consisting of DRL, Cvt, and UDS. Each block represents one of the possible eight sets of positive and negative results of the battery, beginning with three positive results (extreme left) and ending with three negative results (extreme right). (See Table 6.4.1 for the eight combinations.) The ordinate is the predicted carcinogenicity compared to a prior probability of 0.5 and the abscissa is the relative frequency of occurrence.

three additional tests. This information would be useful in determining how much testing is required to offset the positive result in Sty had the result been a false positive.

In order to illustrate this, consider the previous example using Sty, Mly, and SCE (Table 6.4.2 and Fig. 6.4.2) and suppose that we are considering adding Cvt as a fourth test. Of the 16 possible combinations of test results using this four-test battery, only the five test results given in Table 6.4.4 satisfy the requirement that θ^+ is less than 0.1. Note, however, that the likelihood of occurrence of these acceptable results (as computed by the CPBS) is 30%—which is quite high. We also can compute the expected occurrence knowing the Sty result. If Sty is positive and the remaining three tests are all negative, then the value of θ^+ is less than 0.1 and we can compute that the likelihood of this result occurring is 5.5%. If Sty is negative, then there are four combinations of results for the remaining three tests for which $\theta^+ \le 0.1$, and the likelihood of getting these results for the three tests

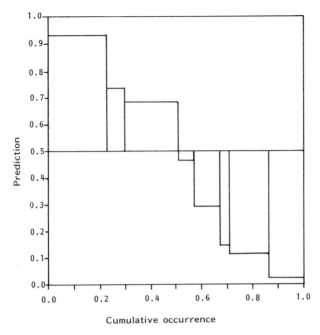

FIGURE 6.4.2. Graphical representation of the predictivity of a battery consisting of SCE, Sty, and Mly. The eight possible results and their θ^+ values are listed in Table 6.4.2. See legend of Fig. 6.4.1 for further description.

is 46.4%. Thus, after using Sty as the first test, there is a 51.9% chance that the results of the remaining tests will be conclusive. In contrast, consider the battery consisting of Sty, Mly, UDS, and Cbm (Table 6.4.4). Note that only when all four tests are negative does the value of θ^+ fall into the acceptable range, and this occurs only with a likelihood of 8.3%.

The preceding analysis can be used in the early stages of product development. This type of analysis not only saves time but also provides a rational way to select tests for battery inclusion. For the specific example given, it has shown unambiguously that the battery consisting of Sty, Mly, SCE, and Cvt would be better (in terms of expected performance) than the battery consisting of Sty, Mly, UDS, and Cbm.

6.5. CLOSING REMARKS

From the illustrations given in this chapter and throughout the book we can see that the CPBS clearly provides an objective approach to the selection and interpretation of multiple test results. It allows one to begin

TABLE 6.4.4. Expected Overall Occurrence of Acceptable Results

Batteries and results				θ^+	Overall percent	Percent after knowing Sty result
A. Sty	Mly	Cvt	SCE			
−	−	−	−	0.0043	9.0	15.1
−	+	−	−	0.0213	9.2	15.3
+	−	−	−	0.0273	2.2	5.5
−	−	−	+	0.0649	4.8	8.0
−	−	+	−	0.0649	4.8	8.0
				Total	30.0	51.9
B. Sty	Mly	UDS	Cbm			
−	−	−	−	0.0244	8.3	14.0
−	+	−	−	0.1128		
−	−	−	+	0.1128		
+	−	−	−	0.1410		
−	−	+	−	0.1410		
				Total	8.3	14.0

with the raw data about the past performances of the tests and to determine the abilities of the tests to identify the properties of interest, and it provides the means for the successful deployment of batteries of tests.

REFERENCES

Ashby, J., 1986, "The prospects for a simplified and internationally harmonized approach to the detection of possible human carcinogens and mutagens," *Mutagenesis*, **1**:3–16.

Chankong, V., Haimes, Y. Y., Rosenkranz, H. S., and Pet-Edwards, J., 1985, "The Carcinogenicity Prediction and Battery Selection (CPBS) method: A Bayesian approach," *Mutation Res.* **153**:135–166.

Dearfield, K. L., Abnerathy, C. O., Ottley, M. S., Brantner, J. H., and Hayes, P. F., 1988, "Acrylamide: Its metabolism, developmental and reproductive effects, genotoxicity, and carcinogenicity," *Mutation Res.*, **195**:45–47.

Ennever, F. K., and Rosenkranz, H. S., 1986a, "Predicting the carcinogenicity of the aromatic amine derivatives tested in the second UKEMS Collaborative Study," *Mutagenesis*, **1**:119–123.

Ennever, F. K., and Rosenkranz, H. S., 1986b, "Short-term test results for NTP noncarcinogenesis: An alternate, more predictive battery," *Environ. Mutagenesis*, **8**:849–865.

Ennever, F. K., and Rosenkranz, H. S., 1986c, "Evaluating batteries of short-term genotoxicity tests," *Mutagenesis*, **1**:293–298.

Ennever, F. K., and Rosenkranz, H. S., 1987a, "Evaluating the potential genotoxic carcinogenicity of methyl isocyanate," *Toxical. Appl. Pharmacol.*, **91**:502–505.

Ennever, F. K., and Rosenkranz, H. S., 1987b, "Selection of batteries in an industrial setting," *Environ. Mutagenesis*, **9**:359–361.

Ennever, F. K., and Rosenkranz, H. S., 1988a, "The influence of the proportions of carcinogens on the cost-effectiveness of short-term tests," *Mutation Res.*, **197**:1–13.

Ennever, F. K., and Rosenkranz, H. S., 1988b, "Methodologies for interpretations of short-term test results which may allow reduction in the use of animals in carcinogenicity testing," *Toxicol. Ind. Health*, **4**:137–149.

Ennever, F. K., Noonan, T. J., and Rosenkranz, H. S., 1987, "The predictivity of animal bioassays and short-term genotoxicity test for carcinogenicity and noncarcinogenicity to humans," *Mutagenesis*, **2**:73–78.

Gold, L. S., de Veciana, M., Backman, G. M., Magaw, R., Lopipero, Pl., Smith, M., Hooper, N. K., Havender, W. R., Bernstein, L., Peto, R., Pike, M. C., and Ames, B. N., 1984, "A carcinogenic potency database of the standardized results of animal bioassays," *Environ. Health Perspect.*, **58**:9–319.

Griesemer, R. A., Harper, C., Calabrese, E., Michalopoulos, G., Rosenkranz, H. S., Schneiderman, M., and Sipes, I. G., 1985, "Report to the U.S. Consumer Product Safety Commission by the Chronic Hazard Advisory Panel on Di-(2-Ethylhexyl)phthalate.

Haimes, Y. Y., Chankong, V., Pet-Edwards, J., and Rosenkranz, H. S., 1987, "Carcinogenicity prediction and battery selection procedure: An in depth analysis of cyclamate and its major metabolite cyclohexylamine," *Molec. Toxicol.*, **1**:49–60.

Havel, R. J., Griesemer, R. A., Lagakos, S. W., Munro, I. C., Pitot, H. C., Rosenkranz, H. S., Stellman, S., Tardiff, R. G., Thomas, D. W., Ward, J. N., Weinhouse, S., and Williams G. M., 1985, *Evaluation of Cyclamates for Carcinogenicity*, National Academy of Sciences Press, Washington D.C.

Hu, S., 1985, "A multiobjective risk methodology for the optimal selection of a battery of tests," M.S. thesis, Case Western Reserve, University.

IARC, 1986, "IARC Monograph on the Evaluation of the Carcinogenic Risk of Chemicals to Humans, Some Chemicals Used in Plastics and Elastomers," Vol. 39, pp. 41–66, International Agency for Research on Cancer, Lyon, France.

IARC, 1988, "IARC Monograph Supplement 7," International Agency for Research on Cancer, Lyon, France.

Jerina, D. M., Lehr, R. E., Yagi, H., Hernandez, O., Dansette, P. M., Wislocki, P. G., Wood, A. W., Chang, R. L., Levin, W., and Conney, A. H., 1976, "Mutagenicity of benzo(a)pyrene derivatives and the description of a quantum mechanical model which predicts the ease of carbonium ion formation from diol epoxides," in F. J. de Serres, J. R. Fouts, J. R. Bend, and R. M. Philpot (eds.): *In Vitro Metabolic Activation in Mutagenesis Testing*, Elsevier/North-Holland Biomedical Press, Amsterdam, pp. 159–178.

Lave, L. B., Ennever, F. K., Rosenkranz, H. S., and Omenn, G. S., 1988, "Information value of the rodent bioassay," *Nature*, **336**:631–633.

Matsushima, T., 1987, *Chemical Safety Evaluation in Japan*, 175.

Miller, J. A., 1970, "Carcinogenesis by chemicals: An overview," *Cancer Res.*, **30**:559–576.

Miller, J. A., and Miller, E. C., 1977, "Ultimate chemical carcinogens as reactive mutagenic electrophiles," in *Origins of Human Cancer*, H. H. Hiatt, J. D. Watson, and J. A. Winsten (eds.), Cold Spring Harbor Laboratory: Cold Spring Harbor, pp. 605–627.

Nesnow, S., Argus, M., Bergman, H., Chu, K., Frith, C., Helmes, T., McGaughey, R., Ray, V., Slaga, T. J., Tennant, R., and Weisburger, E., 1987, "Chemical carcinogens: A review and analysis of the literature of selected chemicals and the establishment of the Gene-Tox Carcinogen Data Base," *Mutation Res.*, **185**:1–195.

Paladja, M., and Rosenkranz, H. S., 1985, "Assembly and preliminary analysis of a genotoxicity data base for predicting carcinogens," *Mutation Res.*, **153**:79–134.

Pet-Edwards, J., 1986, "Selection and interpretation of conditionally independent tests for binary predictions: A Bayesian approach," Ph.D. dissertation, Case Western Reserve University, Cleveland, Ohio.

Pet-Edwards, J., Chankong, V., Rosenkranz, H. S., and Haimes, Y. Y., 1985a, "Application of the carcinogenicity prediction and battery selection (CPBS) methodology to the Gene-Tox data base," *Mutation Res.*, **153**:187–200.

Pet-Edwards, J., Rosenkranz, H. S., Chankong, V., and Haimes, Y. Y., 1985b, "Cluster analysis in predicting the carcinogenicity of chemicals using short-term assays," *Mutation Res.*, **153**:167–185.

Rosenkranz, H. S., and Ennever, F. K., 1987, "Evaluation of the genotoxicity of theobromine and caffeine," *Food Chem. Toxicol.*, **25**:247–251.

Rosenkranz, H. S., and Ennever, F. K., 1988a, "New approaches to battery selection and interpretation," in: *New Trends in Genetic Toxicology*, G. Jolles and A. Cordier (eds.), Academic Press, New York, in press.

Rosenkranz, H. S., and Ennever, F. K., 1988b, "Quantifying genotoxicity and nongenotoxicity," *Mutation Res.*, **205**:59–67.

Rosenkranz, H. S., and Klopman, G., 1987, "Computer automated structure evaluation of the carcinogenicity of N-nitrosothiazolidine 4-carboxylic acid," *Chem. Toxicol.*, **25**:253–256.

Rosenkranz, H. S., Ennever, F. K., Chankong, V., Pet-Edwards, J., and Haimes, Y. Y., 1986a, "An objective approach to the deployment of short-term tests predictive of carcinogenicity," *Cell Biol. Toxicol.*, **2**:425–440.

Rosenkranz, H. S., Frierson, M. R., and Klopman, G., 1986b, "Computer-automated prediction of the mutagenicity of benzidine, 4,4″-diaminoterphenyl, 4-dimethylaminoazobenzene and 4-cyanodimethylaniline: Comparison with the results of the Second UKEMS Collaborative Study," *Mutagenesis*, **1**:275–282.

Rosenkranz, H. S., Frierson, M. R., and Klopman, G., 1988, "Predicting the carcinogenicity of pyrene, benzo(a)pyrene, 2-acetylaminofluorene and 4-acetylaminofluorene using newly developed computer based methods," in: *Evaluation of Short-Term Tests for Carcinogens, Report of the International Programme on Chemical Safety's Collaborative Study on in vivo Assays*, J. Ashby, *et al.* (eds.), Cambridge University Press, in press.

Tennant, R. W., Margolin, B. H., Shelby, M. D., Zeiger, E., Haseman, J. K., Spalding, J., Caspary, W., Resnick, M., Stasiewicz, S., Anderson, B., and Minor, R., 1987, "Prediction of chemical carcinogenicity in rodents from *in vitro* genotoxicity assays," *Science*, **236**:933–941.

Weisburger, J. H., and Williams, G. M., 1981, "Carcinogen testing: Current problems and new approaches," *Science*, **214**:401–407.

Yander, G., Lin, G. H. Y., and Mermelstein, R., 1987, "Selection of batteries in an industrial setting (Letter to the Editor)," *Environ. Mutagenesis*, **9**:357–358.

Zeiger, E., 1987, "Carcinogenicity of mutagens: Predictive capability of the *Salmonella* mutagenesis assay for rodent carcinogenicity," *Cancer Res.*, **47**:1287–1296.

Epilogue

It is a common practice to use multiple tests and sources of information in making risk-based decisions; however, the manner in which the information is integrated and interpreted differs. Well-established statistical methods for interpreting the results of a single test or a single repeated test—tests that are affected by random errors—are available. A large number of methods have been proposed for combining information from various sources, with the field being dominated by statistical decision theory and with a particular emphasis on Bayesian inference. The CPBS approach is grounded on Bayesian theory and provides a systematic mechanism—a synthesis of systems, statistical, and decision analyses—for transferring information from tests and experts into results which can aid decision making.

When using Bayesian methods for combining various sources of information, a problem arises when the sources of information are correlated or dependent. Bayesian methods require estimates for the correlation, covariance, and/or dependence among all sources of information. In the cases where an adequate experimental data base is available, the covariance matrix can easily be estimated. However, in many instances (especially those involving the estimation of environmental health risks), there is a lack of sufficient data. Computationally efficient and convenient methods for identifying and quantifying the biases and dependencies of random variables in Bayesian probability updating procedures are needed. The CPBS approach provides such methods through the development and use of a new measure of dependence—the K_P measure—described in the text.

One of the objectives of the CPBS approach is to improve management decision making by incorporating knowledge about dependencies among sources of information into Bayesian methods. The K_P measure discussed in this book is a measure of dependency that quantifies both the level and direction of dependency. The approach is currently directed to handling sources of information that are binary in nature, i.e., each source of information yields either a positive or a negative contribution. Since test results are often continuous in nature, we plan to extend the theoretical and methodological basis of the K_P measures further to handle continuous tests in Bayesian analyses.

Often the available data base is incomplete, and consequently exact measures of dependence cannot be computed. Management decision making would certainly be improved if additional insight were provided into the uncertainty that is introduced into Bayesian methods due to lack of knowledge about the correlations among the various sources of information. Under conditions of incomplete information, the results presented in this book provide a starting point for quantitatively characterizing the impact of dependence on the various managerial decision options, thus providing the ranges of dependence within which a decision option will remain optimal.

We have seen that a single source of information generally provides inadequate information by which to determine whether a given property is present in a tested object (e.g., a single short-term *in vitro* test on a chemical is inadequate to determine whether it is a carcinogen). However, a battery of multiple tests, when adequately selected, often provides important information for determining whether the given property is present. Although the use of multiple tests can enhance the state of knowledge, it can also greatly increase the complexity of the decision-making process. When more than one test (i.e., a battery of tests) is to be selected in an optimal way, one must (1) consider all possible combinations of tests (i.e., all possible batteries of tests), (2) determine how well each battery is expected to perform, and (3) select the "best" combination of tests (in terms of cost, performance, time, etc.). The approaches for selecting batteries of tests described in this book are generally applicable when the tests are independent, and transform the battery selection problem into a decision-making problem with multiple objectives. The use of tools from the field of multiple criteria decision making (e.g., multiobjective dynamic programming) has enabled us to markedly reduce the number of alternative batteries that should be considered. Combining the statistical dimension of the battery selection problem (namely the dependency among tests) with the multiobjective aspect yields a complex (albeit challenging) research problem. Certainly the material in this book provides an excellent starting point in the pursuit of such research initiatives.

Currently, both the use and the interpretation of short-term test results are in a state of flux due, in a large part, to the recognition of the possible importance of nongenotoxic carcinogens, which are those agents that will be scored as noncarcinogenic by the CPBS method. However, recognition of the possibly confounding effects of such carcinogens has been followed by a realization that such nongenotoxic carcinogens are primarily species, sex, or tissue specific, while genotoxic carcinogens (i.e., those predicted to be carcinogenic by the CPBS method) occur in more than one species and at multiple sites and are not sex specific. It has been further realized that the great majority of human carcinogens are indeed genotoxic, that is, they are predictable by the CPBS method.

At this time, consensus is building that empirical approaches (such as the CPBS method or the one being developed by the International Commission for Protection Against Environmental Mutagens and Carcinogens) are useful for identifying human cancer risks. It has further been suggested that nongenotoxic agents will have to be evaluated in depth on a chemical-by-chemical basis. Finally, it is realized that empirical approaches will have to be paralleled by examination and understanding of the structural basis of the action of cancer-causing agents. Indeed, a variety of efforts are under way to develop methodologies for predicting genotoxic carcinogenicity based upon the recognition of structural features. Obviously, these evaluations can be incorporated into the CPBS methodology as the prior probability. As more genotoxic carcinogens are identified through these efforts, the CPBS methodology will continue to evolve both as an evaluation tool for short-term tests and as a prediction tool for carcinogens.

Appendix

List of Abbreviations for Bioassays and Short-Term Tests

Bfl	Mutagenicity assay using body fluids
BHK	Transformation of BHK21 cells
Bsr	*Bacillus subtilis* repair assay using rec$^+$ and rec$^-$ strains
CA	Animal carcinogenicity
Cbm	Bone marrow cytogenetics *in vivo*
Ch	Human carcinogenicity
CHO	Specific gene mutations in Chinese hamster ovary cells
Cle	*In vivo* leukocyte cytogenetics
Coo	*In vivo* oocyte cytogenetics
Csg	*In vivo* spermatogonia cytogenetics
Csp	*In vivo* spermatocyte cytogenetics
Cvt	Mammalian cytogenetics *in vitro*
C3H	Transformation assay
Dan	*Drosophila melanogaster* aneuploidy
DCM	*Drosophila melanogaster* chromosome mutation
DRL	*Drosophila melanogaster* sex-linked recessive lethal test
Drp	DNA-repair assays under eukaryotic systems
EcW	*E. coli* WP2 reverse mutation assay
HMA	Host-mediated assays
Htr	Heritable translocation in mouse
Mly	Specific gene mutations in mouse lymphoma L5178Y cells
Mnt	Micronucleus test

MPR	Mouse prostate transformation assay
Msl	Mouse specific-locus test
Msp	Mouse spot test
PlA	*E. coli* DNA-repair assay using polA$^+$ and polA$^-$ strains
PrM	*Proteus mirabilis* DNA-repair test
SCE	Sister chromatid exchange
SHE	Transformation of Syrian hamster embryo cells
Spo	Genetic effects in *Schizosaccharomyces pombe*
Sty	Salmonella mutagenicity assay
UDS	Unscheduled DNA synthesis
VET	Viral enhancement systems
V79	Specific gene mutations in V79 Chinese hamster cells
3T3	Balb/c-3T3 neoplastic transformation assay

Index

Accuracy of test, 74
A priori knowledge, 7
   probability (also called prior probability), 8
Assignment potential, 74
Assignment strength, 74

Batch updating, formulation, 24, 166, 172
Battery
   interpretation, 5, 8, 165–173
   selection methodologies, 5, 8, 125–164
      dynamic programming, 148–162
      enumeration, 138–142
      heuritic approach, 142–148
Battery performance: *see* Predictivity, Sensitivity, Specificity
Battery prediction problem, 165; *see also* Battery interpretation
Battery selection problem, 4, 125
Bayes' formula: *see* Bayes' theorem
Bayes' theorem, 8, 16, 22–25, 68, 75, 101, 104, 116, 166
Bayes–Laplace rule, 19
Bayesian-based procedures, 16
Bayesian decision analysis, 13–31, 194–200
Bayesian prediction: *see* Bayes' theorem
Bayesian update: *see* Recursive formula
Bellman's principle of optimality, 54, 148
   example, 59–61

Cancer hazard identification, 4, 8–13
   example, 175–200
CASE methodology, 10
Case clusters, 9, 16
Chi-square statistic, 108–114
Classes of test, 145, 185
Cluster, interpretation of, 41–42
Cluster analysis
   complete link, 39–41, 89
   in CPB, 85–100
   general, 7, 31–43
   improve α^+, α^-, 79–80
   single-link, 37–38, 89, 114, 146
   spanning tree, 38–39, 89, 114
Cluster analysis methodology, 31–32
Cluster analysis techniques, 31
   hierarchical, 34, 36–41
   partitional, 34, 36
Cluster verification, 31, 42–43, 94–100
Complementarity matrix, 87
Complete-link method: *see* Cluster analysis techniques
Conditional dependence
   definition, 115
   K_P measures, 116–121
Confidence intervals
   for dependence measures, 118
   for sensitivities and specificities, 76–79
Criterion, in cluster analysis, 31

Data base, 5
  Gene-Tox, 71, 77, 79, 80, 88, 112, 146
  IARC, 71
  National Toxicology Program, 72, 118
  preliminary analysis of, 8, 80, 88, 121–122
  properties of, 70–74
Data representation, 31, 33–36
  in cluster analysis, 33–36
  in data base, 70–74
Decision analysis: *see* Bayesian decision analysis
Decision nodes, 27
Decision point approach, 10
Decision rule
  in Bayesian analysis, 19, 25
  in computing battery performance, 128–129
  negative consensus rule, 129
  negative majority rule, 129
  negative predictive rule, 129
  positive consensus rule, 128
  positive majority rule, 128
Decision tree, 27–30
Decision variable, 49
  maker, 48
  making process, 16
  making unit, 48
  multiobjective, 3
  tools, 15
Decision making
  under imperfect information, 19
  under no information, 18
  under perfect information, 18
Degree of belief, 17
Dependence, 7
  definition of, 107
  effects of, 101–107
  measures of, 7
    chi-square, 108–114, 146
    K_P measures, 7, 101, 116–121
  See also Conditional dependence
Dominated solution, 140; *see also* Noninferior
Dynamic programming, 17, 53–62, 126

Efficient solutions: *see* Noninferior
Epidemiological techniques, 9
Errors
  random, 4
  systematic, 4
Expected monetary value, 26
Expected utility, 26
Exploratory data analysis, 7, 16, 31, 85–100

False negatives, 12
False positives, 12

Gene-Tox data base: *see* Data base
Genetic spectra, 12

Hierarchical clustering methods: *see* Cluster analysis techniques
Highest effective dose, 12
Hurwitz's rule, 19

Ideal battery, 185
Ideal data base, 7, 73–74
Imperfect information, 19
Inbetweenist rule, 19

Kenny and Raiffa, 26
K_P measures: *see* Dependence measures
Likelihood of carcinogenicy, 4
Logit analysis, 12
Long-term animal bioassays, 9
Lowest effective dose, 12

Marginal probability, 22
Matching coefficient, 35, 87
Measures
  of closeness, 36
  of performance: *see* Sensitivity, Specificity
Multiple-criteria decision making (MCDM), 44
Multiple-objective decision making, 43–53, 16
  optimization, 17

Negative consensus rule, 129
Negative majority rule (see decision rule), 129
Negative predictive rule, 129
Negative proximities, 89
Nondominance bounded, 151
Nondominated: *see* Noninferior solutions
Noninferior solutions, 49, 50, 126, 140, 155

Objects
  in cluster analysis, 31
  in CPBS, 5
Objective probability, 20
Optimal partition of data, 31
Outcome nodes, 27

Pareto optimal: *see* Noninferior solutions
Partional clustering methods: *see* Cluster analysis techniques
Pattern matrix, 33–34

Index

Perfect information, 18
Positive consensus rule, 128
Positive majority rule, 128
Positive predictive rule, 129
Prediction of problem, 4
Predictive value, 74
Predictivity
  of battery, 133–137
  of test, 74
Preference, 17
Preliminary analysis of data base: see Data base
Preposterior analysis, 27
Principle of optimality: see Bellman's principle of optimality
Prior estimate, 20–21
Properties of interest, 4
Proximity level, 37
Proximity matrix, 34–35

Random errors, 4
Recursive formula (equation)
  in Bayesian analysis, 16, 22–25, 168–169, 179
  in dynamic programming, 54, 155
Risk, 3
  avoider, 19
  of carcinogenicity, 3
  of disease, 3
  neutral, 26
  seeker, 19

Selective ability, 7, 74, 75
Semi-ideal battery, 186
Sensitivity
  of battery, 127–133
  of test, 7, 22, 74, 75, 179
Separability, 150

Sequential formulation: see Recursive equations
Serial-monotonic objective function, 151
Short-term tests, 9, 10
Simple matching coefficients: see Matching coefficient
Single-link method: see Cluster analysis
Spanning tree method: see Cluster analysis
Specificity
  of battery, 127–133
  of test, 7, 22, 74, 75, 179
Stages
  of a decision tree, 28
  of dynamic program, 55
State
  of belief, 17
  of dynamic program, 55
  of mind, 17
Structure-toxicology relationship, 9
Subjective probability, 20–21
Systematic errors, 4

Terminal branches, 27
Tests, 5

Utility function, 25

Value elements, 46
Value of perfect information, 30
Value system, 47
Variance of K_P measure, 118
Vector optimization: see Multiobjective optimization

Worth of information, 27